Molecular Biology
Biochemistry and Biophysics
29

Erich Heinz

Mechanics and Energetics of Biological Transport

With 35 Figures

Springer-Verlag Berlin Heidelberg New York 1978

Prof. Dr. ERICH HEINZ
Gustav Embden-Zentrum der Biologischen Chemie
Klinikum der Universität
Theodor Stern-Kai 7
D-6000 Frankfurt 70

Present address:
Cornell University Medical College Dept. of Physiology
York Avenue
New York, N.Y. 10021

ISBN 3-540-08905-5 Springer-Verlag Berlin Heidelberg New York
ISBN 0-387-08905-5 Springer-Verlag New York Heidelberg Berlin

Library of Congress Cataloging in Publication Data. Heinz, Erich, 1912-. Mechanics and energetics of biological transport. (Molecular biology, biochemistry, and biophysics; 29.) Bibliography: p. Includes index. 1. Biological transport. 2. Bioenergetics. I. Title. II. Series. QH509.H44   574.1   78-16106

The use of registered names, trademarks, etc. in this publication does not imply, even in the absence of a specific statement, that such names are exempt from the relevant protective laws and regulations and therefore free for general use.

Offsetprinting and bookbinding: Brühlsche Universitätsdruckerei, Lahn-Gießen.
2131/3130-543210

To my mentor and friend Walter Wilbrandt

Preface

This book deals with energetics of transport processes, largely expressed in terms of the thermodynamics of irreversible processes. Since at the present time too little is known about the molecular mechanism of transport, the present treatment is based largely on hypothetical models. Care has been taken, however, to define the crucial features of these models as generally as possible, so that the equations do not depend too much on hypothetical details. Accordingly, most equations, though developed on the basis of a mobile carrier (ferryboat) model, should apply equally to a conformational model, with an appropriate reinterpretation of the symbols. To better elucidate the essentials, the models are greatly simplified by special assumptions. Maximally, only two flows are assumed to be present in each model at one time: e.g., two solute flows, the flow of solvent and of one solute, the flow of solvent and of heat. The simplifying assumptions may often be unreal. Hence the equations should not be applied uncritically to actual mechanisms. They may at best serve as a basis on which the more appropriate equations may be developed.

The book is not designed to give a complete kinetic analysis of the transport processes described. The kinetic equations are kept to the minimum required to describe the model concerned and to relate it to the corresponding thermodynamic equations. The intention is to stress the close relationship between bioosmotic (transport) and biochemical processes in metabolism. Hence the transport systems and their behavior are treated primarily in terms of biochemistry, using the same terminology and based on the same principles.

Since the primary intention is to give an introduction into basic principles, no attempt has been made to quote the underlying literature completely and according to priority. References are given only to the extent that they seem suitable to help to find more detailed information.

I am greatly indebted to W. Wilbrandt, B. Hess, and P. Geck for their advice and their constructive criticism as to various parts of this book. I would like to express my gratitude to Mrs. E. Kemsley for skillfully typing the manuscript and for her help in its organization, and to Mrs. B. Pfeiffer for the drawing of the figures and technical help in composing the manuscript.

Summer 1978   Erich Heinz

Contents

List of Symbols

A, B		defined solute species
a, b		(molar) concentrations (activities) of A and B, respectively
A_r		affinity of an overall process
A_{ch}		affinity of a chemical reaction
A_{os}		affinity of an osmotic process
a_i		activity of an undefined species i
$\tilde{a}_i$		electrochemical activity of the species i
c_i		(molar) concentration of species i
D_i		Fick diffusion coefficient of species i
F		Faraday constant
G		Gibbs free energy (fee enthalpy)
H		enthalpy
$\bar{H}_i$		partial molar enthalpy of species i
J_i		flux of solute i per unit area
$\overset{o}{\overrightarrow{J}}_i, \overset{o}{\overleftarrow{J}}_i$		initial net flux ("zero trans") of solute i
$\overrightarrow{J}_i, \overleftarrow{J}_i$		unidirectional flux of solute i in the indicated direction
K_i		dissociation constant of carrier complex with solute i
K_m		half-saturation (Michaelis) constant
k		rate coefficient in chemical reaction
L_{ii}, L_{ij}		phenomenological coefficient
L_r		phenomenological coefficient for overall process in quasi-chemical notation
n_i		number of moles of species i
P		hydrostatic pressure
P		product of chemical reaction
P_i		(empirical) permeability coefficient (or "probability of transition") of solute species i
p		concentration (activity) of product P
Q		degree of asymmetry
Q^*		heat of transfer

q	degree of coupling
R	gas constant
r	ratio of affinities of activated to inactive carrier for solute
R_{ii}, R_{ij}	phenomenological resistance coefficient
r_{ii}, r_{ij}	resistance coefficient
S	entropy
S	substrate of biochemical reaction
$\bar{S}_i$	partial molar entropy of substance i
s	concentration (activity) of substrate S
T	absolute temperature
t	time
V	volume
$\bar{V}_i$	partial molar volume of substance i
v	rate of chemical reaction
X_i	osmotic driving force
X, Y	states of transport carrier
x, y	local activities (or probabilities of state) of X and Y, respectively
z_i	number of electric charges of ion i

Greek symbols:

α, β	"normalized" activities of solute species A and B, respectively $[= (a/K_a)$ and (b/K_b), respectively$]$
γ_i	activity coefficient of solute i
Γ	chemical reactivity coefficient $(= e^{\frac{A_{ch}}{RT}})$
δ	thickness of membrane
ε_i	"efficacy of accumulation" of solute species i
η	efficiency
μ_i	chemical potential of solute species i
$\tilde{\mu}_i$	electrochemical potential of solute species i
ν_i	stoichiometric coefficient of solute species i
π	osmotic pressure
ρ	ratio of permeability coefficients (probabilities of transition) of loaded over unloaded carrier through membrane
ψ	electric potential
ζ_i	electrochemical activity coefficient of ion i
σ_i	reflection coefficient of solute i

Abbreviations:

DPG	1,3-diphosphoglycerate
GAP	glyceraldehyde
HR	Haldane ratio
i.c.	infinite cis
LMA	law of mass action
PD	potential difference
3-PG	3-phosphorglycerate
P_i	inorganic phosphate
SPR	sarcoplasmic treticulum
TIP	thermodynamics of irreversible processes
z.t.	zero trans

1 Introduction

1.1 Chemical and Osmotic Processes — A Simple Model

A *chemical* process involves the chemical transformation of one or more solute species into one or more chemically different species. By contrast, an *osmotic* process, in the wider sense, involves the translocation of one or more solute species, through a membrane, usually of limited permeability, without an alteration of the translocated species. Hence, there are two ways a distinct solute can disappear from a given compartment: firstly, by transformation into another species within the same compartment, and secondly, by translocation through a membrane into another, adjacent compartment.

On the same basis the great variety of processes that go on in a living organism and constitute its metabolism can be divided into two equally important groups: the *biochemical* ones and the *bioosmotic* ones. The former refer to chemical reactions in the usual sense, and the latter to the permeation through biological membranes.

Both kinds of processes are closely interlocked within the various metabolic pathways, in which they may occur concomitantly or successively. They depend on each other, or at least influence each other, and may be energetically coupled to each other.

Despite their different nature, biochemical and bioosmotic processes have many common or analogous features, biologically and physiochemically. In either case, if the process goes on continuously, we speak of a flow, for instance of a flow of solute through a membrane, or of a flow of solute, in a figurative sense, "through" a chemical reaction, respectively. In each case, the reaction tends toward an equilibrium, according to the same physicochemical laws. In the biological milieu, both biochemical and bioosmotic processes are largely controlled by specific catalysts, called enzymes in biochemical processes and transport mediators (carriers) in bioosmotic processes, since normally both are prevented from occurring spontaneously by the high activation energy required to form chemical intermediates, or to overcome an osmotic barrier, respectively.

The close analogy as well as the fundamental differences between chemical and osmotic processes may be visualized by two imaginary experiments using the following imaginary model (Fig. 1).

The simple model consists of two well-stirred aqueous compartments, ' and ", separated by a membrane. Initially, two solutes, A and B, are present, e.g., A dissolved in compartment ', and B in

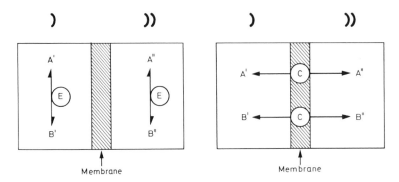

Fig. 1. Difference between chemical and osmotic processes. *A* and *B* stand for different solutes, the superscripts ' and " indicating location in the *left* and *right* compartment, respectively. *Left part*, enzymatic transformation *A* ⟷ *B*, *right side*, carrier-mediated translocation of *A* and *B* between the two compartments. For details see text

compartment ". A and B are interconvertible, but only in the presence of a suitable catalyst (enzyme).

Neither A nor B can penetrate the membrane unless suitable transport carriers are present in the membrane. In the first experiment we start a chemical reaction (A ⟷ B) by adding the required enzyme to either compartment. A will be converted to B in the left compartment, and B to A in the right one, until equilibrium between A and B is established in each. Accordingly in the final state $(a'/b') = (a''/b'')$; but, unless a_o was equal to b_o in the beginning, a' is not equal to a" and b' is not equal to b". In the second experiment we induce an osmotic process. We start from the same initial state, but instead of adding the enzyme we make the membrane permeable to A and B by incorporating a suitable transport carrier. A will now move to the right, and B, to the left compartment, again until an equilibrium is attained. In the final state we have again $(a'/b') = (a''/b'')$. The ratios are not necessarily the same as the corresponding ones obtained after the chemical reaction, but this time a' = a" and b' = b".

We see that the states resulting from the two kinds of processes are similar in one respect and are different in another. These differences are due to the fact that in the chemical reactions equilibrium is reached between A and B in each compartment but not between A' and A" and between B' and B" across the barrier, whereas after the osmotic process the opposite is true.

Complete equilibrium is obtained if both processes can occur together or successively, whereas each of them alone accomplishes only a partial equilibrium. The complete equilibrium is the same, regardless of the order in which the two processes take place.

This model illustrates that both chemical and osmotic processes are equivalent integral parts of the overall process. Owing to this equivalence the two kinds of reactions can be treated in an analogous manner kinetically and energetically.

As stated before, each process, whether chemical or osmotic, tends toward equilibrium. There are two ways to define the equilibrium in either case: kinetically, in terms of the law of mass action (LMA)[1], and energetically, in terms of thermodynamics. By the former definition, equilibrium is reached when the rates of the forward and backward reactions are equal, and by the latter definition, when the change in free energy is the same for the net reaction in either direction. Accordingly both chemical direction and rate of either chemical or osmotic process can be treated quantitatively in two alternative ways: (1) in terms of the LMA, which expresses the rate of the process as a function of the *activities* of the reactive or permeant solutes, respectively; and (2) in terms of the thermodynamics of irreversible processes (TIP), which expresses the rate of the process as a function of the *chemical potentials* of the reactive or permeant solutes, respectively.

1.2 Activities and (Electro)Chemical Potentials

To illustrate these two kinds of treatments, we use again the above model, but instead of looking at the equilibria only, we now try to describe the rate of each the two processes, first in terms of the LMA and then in terms of the TIP. Before doing so we shall briefly characterize the quantitative terms used: the activity and the (electro)chemical potential of the transported solute, respectively, for readers who are not familiar with this topic:

The *activity* of a solute (i) is the product of its molar concentration (c_i) and an activity coefficient

$$a_i = \gamma_i c_i \tag{1.1}$$

γ_i, by definition, is unity at an infinite concentration dilution. At higher concentrations, interactions between the solute particles may affect γ_i. They usually tend to decrease γ_i; however under certain conditions, the may also increase it. A reduction of the activity coefficient may sometimes be simulated by association of the solute particles with each other or with other soluble or insoluble particles. The common neutral organic solutes in biological fluids, such as sugars and neutral amino aicds, occur in concentrations at which γ_i is close to 1 so that their concentrations can safely be used instead of their activities. This does not apply to electrolytes, whose activity coefficients may even in fairly dilute solutions be considerably smaller than unity, especially if the solute carrier more than one electric charge.

---

[1]Setting off the LMA from thermodynamics in this context is merely meant to characterize the formal approach. It is not to imply that LMA is in essence different from thermodynamic laws.

In very dilute solutions, e.g., less than 10 mM, the activity coefficient for a single ion species is a simple function of the "ionic strength" according to the following equation:

$$- \log \gamma_i = 0.5 \cdot z_i^2 \sqrt{\frac{\Sigma z_i^2 c_i}{2}} \qquad (1.2)$$

in which z_i is the valency of the ion, $\frac{\Sigma z_i^2 c_i}{2}$ the ionic strength, i.e., half the sum of the concentrations of all ions present, each multiplied by the square of its valency.

As we shall see later, this more general definition of the activity may for special purposes be extended to include "special" coefficients. Such special activity coefficients may, for instance, be applied to solutes in order to characterize their reactivity in a special chemical reaction, or to ions in order to characterize to their behavior under the influence of an electric field.

The *chemical potential* of a solute (i) is defined as the partial derivative of the total free energy (G) of the system whith respect to the change of n_i, the amount of solute i in the system, while pressure (P), temperature (T), and the amounts of other components (n_j) of the system are kept constant:

$$\mu_i = \left(\frac{\partial G}{\partial n_i}\right)_{P, T, n_j,} \qquad (1.3)$$

μ_i depends on many variables, the most important of which are expressed in the following, well-known differential equation

$$d \mu_i = \left(\frac{\partial \mu_i}{\partial T}\right)_{P, n_i, n_j} dT + \left(\frac{\partial \mu_i}{\partial P}\right)_{T, n_i, n_j} dP + \left(\frac{\partial \mu_i}{\partial n_i}\right)_{P, T, n_j} dn_i + \Sigma \left(\frac{\partial \mu_i}{\partial n_j}\right)_{P, T, n_i} dn_j \qquad (1.4)$$

This equation can be simplified by replacing the partial derivatives $\left(\frac{\partial \mu_i}{\partial T}\right)_{\cdots}$ and $\left(\frac{\partial \mu_i}{\partial P}\right)_{\cdots}$ by the partial molar entropy of i, $\bar{S}_i$, and the partial molar volume of i, $\bar{V}_i$, respectively. Furthermore, $\left(\frac{\partial \mu_i}{\partial n_i}\right)_{\cdots} dn_i$ is RT d ln c_i, $\Sigma \left(\frac{\partial \mu_i}{\partial n_j}\right)_{n_i, P, T} dn_j$ stands for the dependence of μ_i on the components other than i of the system. It is small and often neglected as far as other solutes are concerned, but probably not so if n_j refers to the solvent. The sum of these terms may be interpreted as representing changes of the activity coefficient (γ_i) of i due to changes of the composition of the system. Equation (1.4) can now be simplified as follows:

$$d \mu_i = \bar{S}_i \, d \, T + \bar{V}_i \, d \, P + RT \, d \, \ln \gamma_i \, c_i$$

Dealing with isothermal systems $(dT = 0)$ only, we may drop the first term of the right side. The second term is likely to be very small since $\bar{V}_i$ is usually less than 50 ml/mol (for NaCl it is about 20 ml/mol) and can therefore be neglected in most cases, even in bacteria whose walls permit pressure differences of several atmospheres[2], so that Eq. (1.5) reduces to

$$d \mu_i = RT \ d \ln \gamma_i \ c_i \ = \ RT \ d \ln a_i, \qquad (1.5)$$

since $\gamma_i \ c_i$ can be replaced by a_i, the (chemical) activity of i.

A distinction between concentration and activity is often omitted, i.e., it is assumed that γ_i is close to 1. If γ_i is equal on both sides of the membrane, it cancels out in the derivation of the chemical potential difference of i.

Since the transport of a solute carrying an electric charge, i. e., of an ion, is driven not only by its chemical potential gradient but also by an electric potential gradient, the latter has to be included in the overall driving force. For this purpose the term *electrochemical potential* has been introduced, which by definition is the sum of the chemical and the electric potentials (μ) of the ion i

$$\tilde{\mu}_i = \mu_i + z_i F \psi \ ^{(3}$$

z_i is the electric charge of the ion i, F is the Faraday constant, and ψ the electric potential.

Accordingly

$$d\tilde{\mu}_i = d\mu_i + z_i F d\psi.$$

Expanding Eq. (1.6) accordingly, we obtain

$$d\tilde{\mu}_i = RT \ d \ln a_i + z_i F d\psi . \qquad (1.6)$$

By integration of Eq. (1.6) at constant temperature (T) we obtain

$$\tilde{\mu}_i = \mu_i^o \ (T) + RT \ \ln a_i + z_i \ F \ \psi . \qquad (1.7)$$

$\tilde{\mu}_i$ is called the electrochemical potential because it contains a chemical and an electric term.

---

[2] dP may, however, be important for the bulk flow of the solvent across a membrane. The solvent flow may then also influence the transport of the solute i to the extent that it is dragged along with the solvent. Since this effect, however, does not act through influencing μ_i, it does not belong in this expression.

[3] This term is used for practical convenience only because the electric potential is not a thermodynamically precise quantity.

μ_i^o is the standard potential of the solute i, i.e., the chemical potential of this solute i under the condition that the concentration, or activity, respectively of the solute is unity, that the absolute temperature is T, and that the electric potential is zero. μ_i^o, which depends on the chemical nature of i but not on its concentration or activity, is important for the affinity of chemical reactions, but usually unimportant for osmotic processes, to the extent that the chemical nature of the solute i is not altered by the translocation across the membrane.

Let us now describe the processes induced in our model kinetically, i.e., in terms of the LMA. Starting with the *chemical* process, the reaction A $\rightleftharpoons$ B, we presume that spontaneous transformations of A particles into B particles and of B particles into A particles are going on continuously, according to the probabilities of these transformations. Obviously net transformation of A into B takes place, if the probability that A becomes B is greater than the probability that B becomes A. As follows from the LMA, the probability that a solute species reacts is proportional to its activity. Accordingly we can express the rate of the reaction A $\rightleftharpoons$ B as the difference between the two opposing probabilities or, as they are usually called, unidirectional fluxes:

$$v_a = P_a \cdot a - P_b \cdot b, \qquad (1.8)$$

P_a and P_b being the probability or rate coefficients.

The equilibrium is reached if both probabilities are equal, so that $v = o$. The ratio between product and reactant of the reaction is now equal to the equilibrium constant, K_{eq}, which is obviously

$$K_{eq} = \frac{P_a}{P_b} \qquad (1.9)$$

The *osmotic* process can be treated analogously, except that we are dealing here with the probabilities that the species A moves from one compartment to the other through the membrane, or backward, respectively. For such passive movements, however, we can usually assume that the probability coefficients are the same in either direction so that the rate equation reduces to

$$J_a = P_a (a'-a'') \qquad (1.10)$$

Instead of treating the osmotic process as a special case of a chemical reaction, we could with the same justification proceed the other way around, making the rate equation of the osmotic process the general one and treat the chemical process as a special case of it. In this case we should have to redefine the activities for the chemical reaction, for instance by rewriting Eq. (1.8) as follows

$$v_a = k (K_{eq} \cdot a - b) \qquad (1.11)$$

In this context we could consider K_{eq} a special activity coefficient and call the product $K_{eq} \cdot a$ the "reaction activity" of the species A. This, however, is not customary, in contrast to an analogous procedure applied to the osmotic behaviour of ions under the influence of an electric potential difference (PD) across the membrane.

An electric PD will drive an ion across a membrane even if the activity of the ions is the same on both sides. Also here the flow can be expressed as a function of a difference in activity if we redefine the term activity to include the electric effect. This special activity is called "electrochemical activity", defined as the product of the activity in the former sense and a special coefficient: $\dfrac{a_i \, F\psi}{e^{RT}}$; hence the flow of the cation A^+ would be

$$J_a^+ = P_a \, (a' \, e^{\frac{F\psi'}{RT}} - a'' \, e^{\frac{F\psi''}{RT}}) \qquad (1.12a)$$

Since we usually do not know the absolute potential ψ but at best the difference between the two,

$$\Delta\psi = \psi'' - \psi' \qquad (1.12b)$$

we can rewrite the equation, arbitrarily setting $\psi'' \equiv o$:

$$J_a^+ = P_a^* \, (a' \, e^{-\frac{F\Delta\psi}{RT}} - a'') \qquad (1.12c)$$

or, multiplying both numerator and denominator by $e^{\frac{F\Delta\psi}{2RT}}$

$$J_a^+ = P_a^{**} \, (a' \, e^{-\frac{F\Delta\psi}{2RT}} - a'' \, e^{\frac{F\Delta\psi}{2RT}}). \qquad (1.12d)\,[4]$$

$e^{-\frac{F\Delta\psi}{RT}} = \xi^+$ is called the "electrochemical activity coefficient". Obviously, P_a^* and P_a^{**} are now also dependent on the electric PD, as will be discussed in more detail in a special volume on bioelectric effects.

Let us now turn to the alternative way to describe quantitatively the same processes, namely to TIP: Thermodynamics predicts that each reaction, chemical or osmotic, proceeds spontaneously only if it decreases the total free enthalpy, or increases the total entropy, of the system. In other words, there must be an appropriate *effort* to drive either flow. With a chemical process, this effort is called *affinity*, for our model defined as the loss of free energy when one mole of A is transformed into B. For an osmotic process the driving effort is a *force* in the physical sense, for our model defined as the loss of free energy when one mole of A passes from compartment ' to compartment ". Obviously, the

---

[4]This procedure is tantamount to setting arbitrarily ψ in the middle of the barrier equal zero.

affinity is a *scalar*, and the osmotic force, a *vector*. This difference in tensorial order causes mechanistic problems, especially in coupling processes, as we shall see later. For energetic purposes, however, it can be largely ignored.

The molar free energy of a solute species is called the chemical potential (μ_i) (or electrochemical potential in the case of ions, $\hat{\mu}_i$) of this species. Accordingly the affinity (A_{ch}) that drives the chemical reactions in our model is defined as the (negative) difference between the electrochemical potentials of A and B:

$$A_{ch} = - (\mu_b - \mu_a) = \mu_a{}^o - \mu_b{}^o + RT \ln a - RT \ln b. \qquad (1.13a)$$

The electric terms cancel, even if A and B are ions, since within the same compartment the electric potential is the same for both A and B,

$$A_{ch} = - \left[(\mu_b{}^o - \mu_a{}^o) + RT \ln \frac{b}{a} \right] \qquad (1.13b)$$

or, since $(\mu_b{}^o - \mu_a{}^o) = - RT \ln K_{eq}$

$$A_{ch} = RT \ln \frac{b}{K_{eq} \cdot a} \qquad (1.13c)$$

On the other hand, for the osmotic flows of each A and B from one compartment to the other, we can express the driving efforts, or forces (X_a) as

$$X_a = - \Delta \mu_a = - (RT \ln \frac{a''}{a'} + z_a F \Delta \psi) \qquad (1.14a)$$

$$X_b = - \Delta \mu_b = - (RT \ln \frac{b''}{b'} + z_b F \Delta \psi) \qquad (1.14b)$$

This time the standard potentials cancel since each remains the same during the translocation. The electric PD, however, does not necessarily cancel here

$$\text{or} \quad \Delta \mu_a = RT \ln \frac{a''}{a'} - RT \ln e^{-\frac{z_a F \Delta \psi}{RT}} \qquad (1.15a)$$

$$\text{or} \quad \Delta \mu_a = RT \ln \frac{a''}{a' \xi^{z_a}} \qquad (1.15b)$$

The electric PD becomes effective if z_a is different from o, i.e., if A is an ion. For both chemical and osmotic processes we can express the driving effort in terms of the logarithm of an activity ratio, if we accept the factors K_{eq} and ξ^{z_a} as representing special activity coefficients.

One-Flow Systems –
Uncoupled Transport

2 Nonmediated (Free) Diffusion

2.1 Mechanistic Aspect

2.1.1 The Barrier. The mechanism by which a solute is translocated across a biological membrane is still a matter of dispute. In the last decades considerable insight into the chemical composition and the topographical structure of biological membranes has been gained. Moreover, artificial membranes are now available to serve as suitable models to study special features of biological membranes. Still, the connections between morphological features and transport behavior of biological membranes are scarce, so that exploration of transport mechanisms still depends heavily on dynamic observations, such as kinetics, etc.

As to their permeability, biological membranes largely behave like hydrophobic barriers with a certain amount of preformed hydrophilic holes (pores) of a distinct diameter. Accordingly, lipophilic solutes should penetrate a membrane by dissolving in the hydrophobic layer, whereas a hydrophilic solute should diffuse across the pores to the extent that its size, shape, or electric charge permit it to do so. In the present context we shall deal mainly with the penetration of hydrophilic solutes. The question arises concerning the extent to which the simplified concept of aqueous pores as the major pathway for these solutes is justified (DAINTY and HOUSE, 1966).

Available morphological evidence has been plausibly interpreted by SINGER and NICOLSEN (1972) to indicate that biological membrane consist of a continuous layer of liquid hydrophobic (lipid) material, in which proteins are floating (Fig. 2). Some of these proteins are larger than the thickness of the lipid layer, which is estimated to be between 30 and 40 Å, and are located in such a way that they span it. This concept of a lipid layer is easily compatible with the observations on the penetration of lipophilic solvents. On the other hand, pores, as they are postulated on the basis of kinetic observations, have never been unequivocally ascertained in the lipid layer. Possibly the above-mentioned proteins with their hydrophilic meshwork of peptide chains or the interspaces between monomers of dimer proteins account for the phenomena attributed to preformed pores. Also alternative pathways have been considered for the penetration of hydrophilic solutes, for instance directly through the lipid layer, which is considered to be in a continuously fluctuating motion, thereby opening and closing transient holes and pockets to take up and carry solutes from the adjacent solutes (LIEB and STEIN, 1971).

For each of these pathways, via water-filled pores and via the lipid layer, favorable evidence can be quoted, as if in reality both existed side by side:

Fig. 2. Schematic representation of membrane structure. Proteins are represented as elliptic bodies floating in a lipid double layer. (a) Proteins
located at or near either membrane surface. (b) Proteins that partly penetrate the membrane. (c) Proteins that bridge the membrane completely, singly
or in pairs

2.1.2 Diffusion Through Pores. The following two arguments are often
raised in favor of preformed aqueous pores:

1. Under the influence of hydrostatic or osmotic pressure differences water or solution penetrates the membrane many times
faster than could be accounted for by molecular diffusion of water. It has been followed that water in this case passes the
membrane as a continuum (bulk flow). It is difficult to visualize bulk flow otherwise than as passing through preformed pores.
This argument is no absolute proof since similar phenomena have
been observed with nonporous artificial membranes, and there
attributed to errors caused by unstirred layers.

2. Electrolyte ions, under the influence of electric forces,
penetrate biological membranes faster than could be expected
from their extremely low solubility in the membrane lipids. Such
ions, in order to enter a lipid phase, apparently have to strip
off their hydration shell, as appears to follow from the extremely low partition coefficient between a bulk lipid phase and water
in the test tube. Accordingly the actual permeability of biological membranes to monovalent electrolyte ions can hardly be accounted for otherwise than by water-filled pores. This argument
is strong but not fully conclusive either, to the extent that
specific transport mediators, such as will be discussed in a subsequent chapter, cannot always be excluded.

2.1.3 Diffusion Through Lipid Layer. Even if true pores did exist and
could be held responsible for the bulk movement of water and the

nonspecific penetration of electrolyte ions, it would not follow
that organic hydrophilic solutes also use such pores as a major
pathway.

On the contrary, many features of this transport appear to argue
in favor of the alternative pathway, which leads them directly
through the *lipid layer*. The arguments in support of this view,
have been collected and are discussed extensively by LIEB and
STEIN (1971). They are based on some kinetic peculiarities of
organic solute transport that appear to be characteristic of
diffusion within phases of high polymers rather than in water.
According to these authors, the lipid layer of a cellular membrane behaves in this respect like a polymeric medium, at least
like a medium composed of solvent particles that are much larger
than the penetrating solute particles. The penetration of such
media by small molecular shows characteristic kinetic differences
from that of aqueous media, in which the solvent particles are
smaller than the solute particles. For instance, the decrease
in mobility with increasing molecular weight of the solute particles is much more pronounced in high-polymer solvents than in
water. In biological membranes the corresponding dependence of
the permeability on particle size is reported as "steep" as in
high-polymer solutions. Furthermore, particles of equal size diffuse in water the faster the more their shape resembles a sphere,
whereas in high polymeres as well as in biological membranes
spherical shape appears to be detrimental to penetration. This
difference appears to be enhanced by the fact that lipids in
the membrane layer are in a more ordered state than are water
molecules in the bulk phase of water.

Furthermore, raising the temperature accelerates the penetration
of large particles more than that of smaller ones in biological
membranes, whereas in free aqueous solution temperature affects
the mobility of solutes independently of particle size. Finally,
the addition of plasticizers, i.e., of certain low-molecular-
weight solutes, to high polymere phases as well as to biological
membranes, tends to increase the permeability for other solutes,
whereas such plasticizers have little effect on the diffusion of
the same solutes in water.

The kinetics of the passage of hydrophilic solutes directly
through the lipid layer is complicated by the fact that three
distinct steps in series are involved.

1. Entrance of the lipid phase from the water phase

2. Diffusion within the lipid phase

3. Exit from the lipid phase into the water phase of the trans-
   compartment

Whether and which one of these three steps limits the rate of
the overall process may depend on the chemical nature of the solute and is difficult to predict. Apparently, a hydrophilic solute has to disconnect its H bridges before entering the lipid
phase, as there is an inverse relationship between the rate of
penetration and the number of H bridges a compound forms in water. A fairly linear relationship between penetration rate and
concentration difference may still be found, so that also here

the term free diffusion would be qualitatively justified. The apparent diffusion coefficient, however, may not show any predictable relationship to that holding for free solution at all.

So it seems that hydrophilic solutes other than electrolyte ions use the same pathway as do lipophilic solutes, but unlike these they probably do not dissolve in the lipid phase but are being taken up and carried through transient holes and pockets of varying sizes, which are randomly opening and closing within the fluctuating network of polymere-like membrane lipids. On the other hand, it seems that pores, if they exist at all, may serve as the main pathway for water and electrolyte ions, but at best as a minor pathway for small hydrophilic organic solutes. There is a tendency to believe that these pores are provided by proteins within the lipid phase, especially those that are large enough to span the lipid layer, and that the meshes within the network of helical structures and the interspaces in protein dimers might provide the apertures accounting for porous phenomena. From what we know of protein structure, such apertures should not be much wider than about 4 - 5 Å in diameter or somewhat less, but close to previous estimates from water-flow studies (PAGANELLI and SOLOMON, 1957; SOLOMON, 1960). Whether channels of this width can still be regarded as nonspecific pores is questionable.

The nonmediated movement of solutes through biological membranes is called "free diffusion" by many biologists, on the assumption that it is through aqueous pores and that specific carriers or channels are not involved. This assumption is usually based on the observation that the rate of penetration is not saturable, i.e., approximately proportional to the difference in concentration, and is not inhibited competitively by chemical analogs. It is not justified, however, to expect any quantitative relationship between diffusion in free solution and through biological membranes. Even if there are aqueous pores of sufficent width, solutes will probably move through the porous lumen more slowly than they would diffuse through an "equivalent" volume element, i.e., one of similar size and shape, in free solution. As could be demonstrated with artificial membranes that contained pores of cylindrical shape and of a diameter ten times greater than that of the penetrating solute particles, the apparent diffusion coefficient, relative to the available pore area, is very much smaller than the corresponding coefficient for free solution (van BRUGGEN et al., 1974). The deviations are attributed mainly to frictional interaction between the passing solute and the lining of the pores. This friction can hardly be treated like ordinary mechanical friction because the "lining" of the pores is probably neither rigid nor indifferent. It can instead be expected that the filaments of this meshwork has more or less specific sites to combine with some of the passing solutes. These sites may not be fixed in space but may fluctuate more or less irregulary. For the kind of diffusion across biological membranes, in view of the many restrictions mentioned above, the term "restricted diffusion" has been suggested.

A special situation may arise if the width of the pores becomes so small that only one particle can pass through it at a time.

In this case we speak of "single-file diffusion", in which direct interaction between passing particles is predominant (see Sec. 3.2.5).

The various interactions between solute and membrane (solute-membrane interaction) should be expected to retard the penetration of the particle, the more so the smaller the width of the pore relative to the diameter of the passing solute and the greater the tortuosity of the pore. As we shall see later, however, this need not always be so. In so-called *channels*, which are porous structures of a special kind, such interaction may have just the opposite effect, namely, to accelerate the penetration. These channels, unlike the pores discussed in this chapter, can be considered transport mediators and will accordingly be dealt with in Chapter 3.

Additional deviations from simple diffusion may be due to interaction with other solute particles within the pore (solute-solute interaction) or with the solvent (solute-solvent interaction), which is unlikely to have the same viscosity in these meshworks as in free solution.

2.2 Treatment of Unmediated Diffusion in Terms of the Law of Mass Action (LMA)

2.2.1 Diffusion Coefficient (D_i). For free diffusion in free solution Fick's first law applies

$$\mathrm{d}n_i/\mathrm{d}t = -D_i\ A \cdot (\mathrm{d}c\ /\mathrm{d}x) \tag{2.1}$$

It gives the number of moles of solute i (n_i) passing through a two-dimensional plane of the area A. c_i is the concentration of the solute, x the length of the diffusion pathway in the direction of flow. D_i, the so-called diffusion coefficient, is constant for a given solute in a given solution at constant temperature. It depends on physical properties of the solute species, such as particle size and shape, and on the viscosity of the solvent.

For large spherical solutes with the radius r, D_i is approximately inversely proportional to the volume of the solute, i.e., to r^3, in the same solvent. For smaller particles, of a size close to that of the H_2O molecule, D_i is proportional to the area, i.e., to r^2. For the same solute in different solvents, D_i is inversely proportional to the viscosity of that solvent (LIEB and STEIN, 1974).

If the membrane were homogeneous, the Fick equation could, in the steady state, be integrated to

$$\frac{\mathrm{d}n_i}{\mathrm{d}t} = \frac{D_i\ A}{\delta}\ \kappa\ (c_i' - c_i'') \tag{2.2}$$

κ being the partition coefficient of the solute between the membrane phase and the adjacent bulk phases, and δ the thickness of the membrane. It might apply to the penetration of a membrane by lipophilic solutes to the extent that matrix behaves approximately like a liquid phase. It is implied that the overall penetration rate is determined by the diffusion within the membrane phase, whereas the transfer of the solute through the boundary between this phase and the aqueous solutions is instantaneous, and that unstirred layers near the membrane phase are insignificant[5].

Attempts have been made to apply Eq. (2.2) to the penetration of biological membranes by hydrophilic solutes. A was to stand for the total area of aqueous pores per unit area of the whole membrane and has been estimated under the assumption that D_i in the pores is the same as in free solution. This assumption would be permitted only if the pores were of cylindrical shape and so wide that the passing solute could diffuse in them as freely as in free solution. As discussed in Chapter 1, pores in biological membranes, if they exist at all, are far from meeting these requirements, let alone the possibility that many solutes would not even use pores as their main pathway but work their way directly through the lipid layer. In addition, we do not know precisely the thickness of the membrane over which to integrate. Thus, this procedure can only yield approximate values.

2.2.2 Permeability Coefficient (P_i). In view of these uncertainties with biological membranes it has become customary for practical purposes to use the empirical "permeability coefficient" (P_i) instead of D_i. The resulting permeability equation would then be

$$\frac{dn_i}{dt} = P_i\, A\, \Delta c_i, \quad \text{or} \quad J_i = -P_i\, \Delta c_i \qquad (2.3)$$

J_i is used here to designate the "density" of solute flow, i.e., the amount of solute (in mole) penetrating the membrane per unit of reference, e.g., per unit of area.

The introduction of P_i permits the "quasi-chemical" treatment of biological permeation, i.e., analogously to a chemical reaction, as will be carried out below.

The derivations of the kinetic equations are based on a simple model similar to that used in the Introduction. It consists of two compartments, ' and ", separated by a "permselective" membrane, here a membrane permeable only to the species under consideration, A. In order to have a true "one-component" system, A should be present in the compartments as an ideal gas: a solution would not be a truly one-component system because the presence of a solvent cannot be neglected in the treatment of the diffusion of A. If A were a solute in water, its movement through

---

[5]Unstirred layers, which depend on the surface structure of the membrane, but also on the surface charge and ionic strength, may be 50 Å thick, or thicker, can indeed be neglected, at least with single-cell suspensions, since the free diffusion of most solutes through such a layer is very fast compared to transport through the membrane (MILLER, 1972).

the membrane would interact with the concomitant movement of water molecules, especially if it proceeds through aqueous pores. Such interaction can, however, be largely reduced by preventing net movement of water. Most ideal would be a membrane permeable to A but not to water. A more realistic approach would be to enclose the compartments by rigid walls, including the separating membrane. In such a system it could not be avoided that under certain conditions a hydrostatic pressure difference might develop between the compartments. However, such hydrostatic pressures have negligible influence on the activity of the solutes. In addition the following simplifying assumptions have been made:

1. The membrane phase is small compared to the bulk phase so that the amount of penetrating solute present within the membrane phase can be neglected compared to that in the bulk phases at any time.

2. The membrane is in the steady state with respect to penetrating solute, i.e., the amount of this solute entering the membrane phase is always equal to that which leaves it during the same time.

3. The penetrating solute particles do not appreciably interfere with each other.

4. The concentration of the solute in each bulk phase is about the same throughout the whole phase. This condition requires that the penetration of the membrane is slow compared to the movement within the bulk phase. Such is the case if the bulk phases are either well-stirred or so small that considerable activity gradients of the solute concerned cannot be maintained.

2.2.3 Unidirectional Fluxes – The Flux Ratio. In analogy to a chemical reaction, the net permeation of a solute can be divided into two opposing unidirectional fluxes each of which is proportional to the activity (a) of the solute i in the compartment of origin (cis). Accordingly for the penetration of solute A we could write

$$J_i^{net} = \overrightarrow{J_i} - \overleftarrow{J_i} \tag{2.4}$$

$$\overrightarrow{J_i} = P_i a_i' \tag{2.5a}$$

$$\overleftarrow{J_i} = P_i a_i'' \tag{2.5b}$$

P_i would correspond to the rate coefficients of the chemical reaction with the difference that P_i in unmediated transport is identical for both fluxes.

The use of P_i does not commit us to a distinct mechanism of permeation. If the solute, as has previously been assumed, penetrated exclusively by free diffusion along a continuous concentration gradient through aqueous pores, P_i could be considered an *integral* diffusion coefficient, integrated over the thickness (δ) of the membrane under steady-state condition, i.e., assuming that the net flow per unit area is the same for all layers of the membrane

$$\frac{1}{P_i} = \int_0^\delta \frac{dx}{D_i} \tag{2.6a}$$

For a homogeneous membrane phase, in which D_i is constant in the perpendicular direction,

$$P_i = \frac{D_i}{\delta} .$$
(2.6b)

To the extent, however, that the solute moves directly through the lipid layer (see Chap. 1), the overall permeation would be more appropriately envisaged as occurring by discrete "jumps" of the solute over a major energy barrier. In that case P_i could be interpreted as a probability coefficient (S. 3.2).

The permeability coefficient P_i, though being more useful than D_i for practical purposes, can in contrast to the latter be treated as a constant only at steady state and for a given membrane of constant thickness. It should therefore be determined empirically for each membrane to be investigated. Whereas the dimension of D_i is always $cm^2 \ s^{-1}$, the dimension of P_i depends on the frame of reference. The most rationalistic reference would be the area of the penetrated membrane. In this case the dimension of P_i would be $cm \ s^{-1}$. In many cases, however, the area of a biological membrane cannot be precisely determined, especially not for cells suspended in a medium, unless the cells are spheres of known size. Very often, cellular shape is rather irregular and hence difficult to characterize stereometrically. In addition, the cell surface may not be smooth; indentations or microvilli may greatly expand the effective area in an unpredictable way. For these reasons, it has become customary in many cases to use other references, for example, dry weight of the cellular mass, assuming that under normal conditions there is a fixed proportion between the surface area of the cell and its dry weight.

P would here have the dimension $ml \ g^{-1} \ s^{-1}$. In other cases, especially with erythrocytes, one prefers the volume of the cell water as reference. One should, however, keep in mind that the water content of cells is highly variable, depending on the osmolarity of nonpermeant solutes in the suspending medium. Swelling or shrinking of cells may greatly falsify the permeability data, feigning transport processes in cases where there are none. The volume of cell water can therefore be used as reference only if reduced to a *standard value*, usually taken to be the volume of the cellular mass under normal condition of osmolarity, etc. If during the experiment the cell taken up or looses water, the measured cell volume has to be corrected accordingly, before the uptake or loss of solutes can be determined. Under these conditions, P_i has the unit s^{-1}.

Table 1. Dimensions of Permeability Coefficient

Reference	Dimension
Area	$cm \ s^{-1}$
Mass (dry weight)	$ml \ g^{-1} \ s^{-1}$
Volume (standard)	s^{-1}

Another way to overcome uncertainties concerning permeability properties within the membrane is to use the *flux ratio*, i.e., the ratio of the isotopically determined unidirectional fluxes. A unidirectional flux of a solute is obtained by adding a labeled isotope of the solute in tracer amounts to one side of the membrane and by determining the initial rate at which this label appears at the other side. It indicates the rate at which particles of A from the left side pass through the membrane. The opposing flux can be obtained analogously, or, since the net flux is the difference between the two unidirectional fluxes, by subtracting the net flux from the unidirectional flux as determined above

$$\overrightarrow{J}_i - \overleftarrow{J}_i = J_i^{net}$$

USSING (1949) has derived that for free diffusion this ratio is equal to the ratio of activities of the solute concerned at both sides of the barrier

$$\boldsymbol{f}_i = \overrightarrow{J}_i/\overleftarrow{J}_i = \frac{a_i'}{a_i''} \qquad (2.7a)$$

For ions this ratio has to be replaced by that of the electro-chemical activities

$$\boldsymbol{f}_{ion} = \frac{\tilde{a}_i'}{\tilde{a}_i''} = \frac{a_i'}{a_i''} \cdot \exp\{-\frac{F\Delta\psi}{RT}\} \qquad (2.7b)$$

The permeability properties of the membrane cancel each other since they appear in both numerator and denominator.

This flux ratio clearly should differ from the electrochemical activity ratio if the flow is coupled to any other process, be it to the flow of another solute, to that of the solvent, or to the advancement of a chemical reaction, such as in active transport. For this reason, deviations of the flux ratio from the activity ratio were often taken to indicate active transport or solvent drag, in any case, the presence of nonconjugate driving forces, as will be dealt with in the next section. It is true that the validity of the flux equation can thus be taken to exclude coupling to nonconjugate processes such as in active transport, but the converse is not necessarily true: Also in the absence of nonconjugate driving forces, the flux ratio may deviate considerably from the activity ratio. Such deviations may occur whenever the particles of the penetrating solute *interact* with each other, be it directly, or indirectly, via a mediator. This interaction will be detectable with the fluxes, as measured by different isotopes of the same solute, hence the term isotope interaction or tracer coupling.

2.2.4 Electric Driving Forces of Diffusion. Electric PDs across the biological membrane may effect transport processes through this membrane in various ways. Firstly, the electric PD adds to the driving force of the transport of ions and of ion-linked transport of nonelectrolyte solutes. Secondly, electric potentials may also affect membrane structure, for instance by rearranging ionic or dipolar elements of the membrane, thereby opening or closing "channels" or by otherwise changing the permeability properties of this membrane. Such structural effects may depend not only on the magnitude of the potential but also on the orientation. Rectification phenomena are sometimes the result of such effects.

In the present context we shall not deal with such structural effects, since there is too little known about them and since they may differ greatly from membrane to membrane. We shall therefore here deal with the electric potentials as a *driving force* only.

Thermodynamically an electric PD as a driving force is completely equivalent to any other driving force, provided that the driven particle carries an electric charge. It is therefore customary to add the electric PD to the chemical PD, so that both can be combined to a single driving force, linked to the movement of the driven species by a single coefficient. This has been discussed before and little has to be added in the present context.

Somewhat more complicated is the kinetic treatment of electric potential in terms of the LMA. In this treatment, it may be recalled, the flow of a given substrate is taken to be proportional to the difference of concentration or activity. The implementation of an electric PD into a kinetic equation based on the LMA is not simple and depends on mechanistic details of the transport systems. Some simplification is obtained by the *constant field approach* (GOLDMAN), i.e., by assuming that the gradient of the electric potential across the membrane is linear. This is an approximation which, however, appears to be permissible in many cases. Based on this approximation the following equation can be derived by a simple method, and is widely applied. Some variations exist as to the reference potential, which is taken to be zero in this treatment. Often the electric potential on the trans side is used as the zero reference, so that the electrochemical activity coefficient appears only in the left-side term:

$$J_i^+ = P_i \frac{z_i F \Delta \psi}{RT(1-\xi)} (c_i'\xi - c_i'') \qquad (2.8a)$$

z_i is the electrovalency of the ion i. ξ_i the electrochemical activity coefficient ($\exp\{-\frac{z_i F \Delta \psi}{RT}\}$). Other researchers prefer to set some potential in the middle of the membrane phase equal to zero so that a modified electrochemical activity coefficient appears at both the left- and the right-side activity term.

$$\text{or } J_i^+ = P_i \frac{z_i F \Delta \psi}{RT(\xi^{-1/2} - \xi^{1/2})} (c_i'\xi^{1/2} - c_i''\xi^{-1/2}) \qquad (2.8b)$$

It is noteworthy that also the overall permeability coefficient becomes a function of the electric PD, which complicates the treatment. It should be pointed out, however, that this treatment applies only to a homogeneous membrane phase, but nut necessarily

to a biological or artifical membrane, which supposedly contains
a hydrocarbon core as the major permeability barrier. If one assumes that this nonelectric barrier is great compared to the elec
tric PD, treatment may be greatly simplified. In this case the
rate of penetration is a function of the fraction of particles
having enough activation energy to jump across this barrier (HALL
et al., 1973). This fraction can be expressed by the Boltzmann
equation and, provided that the above assumption is valid, simply
added to this activation energy; in other words, PD increases
the number of molecules having this activation energy. It is assumed that in such a case the electric potential does not increase the rates at which a single particle jumps the barrier,
as it would do in a homogeneous phase, but only increases the
number of ions jumping this barrier per time unit. Under these
conditions, the implementation of the electric activity coefficients can be greatly simplified so that the following expression
may be sufficient to describe the process adequately, the overall
permeability coefficient being no longer a function of the electric PD:

$$J_i = P_i' c_i \xi^{1/2} - P_i'' c_i'' \xi^{-1/2} \qquad (2.9)$$

P_i' and P_i'' are different if the system is structurally asymmetric.

3 Mediated ("Facilitaded") Diffusion

3.1 Mechanistic Aspects of Mediation

3.1.1 General. In Chapter 2 it was mentioned that the interaction between diffusing solute particles with components of the membrane may retard the movement of the solute owing to frictional forces. In the present chapter we shall discuss how such interactions may, under special conditions, greatly accelerate the penetration. Apparently such interaction involves a specific component of the membrane, which in essence functions as a "mediator" of the translocation. Also the observation that certain solutes, whose size should exceed the limits set by the membrane structure for free or restricted diffusion, may penetrate biological membranes at great ease, is attributed to "mediation". The translocation that is accelerated or made possible by the interaction with a mediating component of the membrane, is usually called "facilitated diffusion". Mediation by membrane components may also be the basis for coupling the penetration to the other processes, as in active transport etc. In the present context, however, we shall not yet deal with coupling phenomena, rather leave them for a special chapter.

Mediated transport is assumed to involve the transient binding of the penetrating particle to a specific site of a mobile or fixed component of the membrane. This view is based primarily on the following criteria, which are considered to point to such an interaction with a mediator, also called a translocator. Some of these bear a striking resemblance to those of enzymatic reactions:

1. Specificity with respect to the transported substrate

2. Saturation kinetics, frequently fitting the simple Michaelis-Menten equation

3. Competetive and noncompetetive inhibition by substrate analogs, and enzyme poisons, respectively

4. Activation by certain ions and other cofactors and cosubstrates.

5. Genetic and feedback regulation by inducers, hormones, metabolic intermediates of the solute transported and other agents

In addition, there are some criteria that may be considered typical for membrane transport, such as the so-called *trans effects*, i.e., the dependence of the penetration of a given substance on the nature and concentration of solutes present in the phase adjacent to the opposite side of the membrane. Such trans effects are in most cases stimulatory (trans stimulation), but instances of trans inhibition also have been described.

In view of the preceding criteria, it is tempting to regard biological transport mediators as enzymes or enzyme-like effectors with special functions: Whereas substrates selectively bound to an enzyme will undergo a chemical transformation, the corresponding substrates bound to a transport system will undergo translocation across a biological barrier, usually without a chemical transformation. The difference between enzymatic reactions and mediated transport is not as fundamental as it may appear, and in some cases it is difficult to separate them from each other. So also in a chemical enzyme-catalyzed reaction a substrate may undergo a certain translocation, "across the enzyme, or along its surface" (MITCHELL, 1960), so that this substrate after its chemical transformation may be released from the enzyme at a locus different from where it was first bound. Such "microscopic" translocation, however, does not appear as long as the enzyme is randomly distributed in the medium. It may, on the other hand, have macroscopic consequences if each enzyme molecule is specifically oriented inside a barrier. There are special transport systems based on this principle like those of group translocation, as will be discussed in the corresponding chapter. On the other hand, also the translocation by a transport system may involve an at least transient chemical transformation of the transported species, even if in most such systems no such transformation can be detected.

Whatever the nature of a mediator, it has to be endowed with the following capabilities:

1. To select, i.e., to recognize and capture, the appropriate substrate. This capability is the basis of the above-mentioned *specificity*.

2. To enact, or, at least, accelerate, *translocation* of the selected substrate across the osmotic barrier.

3.1.2 Specificity. Transport systems usually select their "substrate" according to a few crucial criteria, as do enzymes. As a rule, transport systems appear to be somewhat less specific than comparable enzymatic systems, i.e., their capability to distinguish between similar solutes seems less pronounced in transport systems. For instance, the transport system for the uptake of glucose accepts more glucose analogs than does the hexokinase, the first enzyme to act upon the sugar taken up by the cell. Accordingly, several members of a given family of substrates may be transported by the same transport system, whereas each member may require a special enzyme for the first metabolic reaction it is to undergo. Similarly, the stereospecificity, i.e., the ability to distinguish between optical isomers of the same solute, is usually complete or nearly complete with enzyme systems but often incomplete with transport systems: For example the physiological L-amino acids are preferred by most cellular transport mechanisms, but their D isomers are usually not entirely rejected. Nevertheless, the principles underlying the recognition step are considered to be the same for both transport systems and enzymes. In either case the selected solute is supposed to be bound, or attached to a grouping of specific sites, probably in most cases by noncovalent bonds. The nature of this selective attachment, unclear as it is at the present moment, is therefore no special transport problem. All views concerning this process in transport mechanism are usually borrowed from enzymology. For this reason they need not be treated extensively in the present context. The reader is referred to a textbook on enzymology.

A special problem arises if the specificity of *ion* transport is to be explained, in particular the varying specificities of carrier or channels for the ions of strong electrolytes, especially Na^+ and K^+. For this kind of specificity there are no counterparts in enzymatic reactions, since these ions are not known to enter any enzymatically catalyzed biochemical reaction via the formation of covalent bonds. Yet there are many biological transport systems, which, like some ionophores in articifial membranes, strongly discriminate between different alkali ions, giving up to more than a hundredfold preference to one species as compared to another one. The order of preference with respect to alkali ions, for instance, varies from system to system. Some systems, for example, prefer Na^+ strongly to K^+, whereas others do the opposite. Li^+ may replace Na^+ in some systems, in others not. The discrimination between K^+ and Rb^+ is usually less pronounced in most systems, but in some it is also noticeable.

The basis of the discrimination between the various ion species has been puzzling researchers for a long time. The earlier view was that in passive permeability the distinction was based on the size of the ion, because the ion-conducting pores were assumed to be so small as to exclude ions above a certain limiting size. One of the difficulties with this view was to decide whether this limiting size referred to the naked ion or to that of the hydrated ion. As can be seen from Table 2, the diameter of the naked Na ion is considerably smaller than that of the naked K ion, whereas for the hydrated ions the opposite is true. With either alternative, taken to hold throughout, the order of preference should be always the same in all systems. It has already been mentioned, however, that this is not the case.

Table 2. Ionic radii of alkali metal ions

	Crystal radius, Å	Hydrated radius, Å from mobility
Li	0.60	2.31
Na	0.95	1.78
K	1.33	1.22
Rb	1.48	1.18
Cs	1.69	1.16
NH₄	1.48	1.21

More recent hypotheses tend to explain these specificities in
terms of electrostatic forces only, assuming that the ions, in
order to be transmitted, have to be bound to a specific site,
be it on a carrier or on a channel, in a more or less dehydrated
form. The binding process is presumably fast enough to establish
almost complete equilibrium between the free and bound ions, before the translocation. Apparently the rate of transport is considered to increase with the stability of binding; this idea is
probably based on the assumption that for a dehydrated ion to
slide from one binding site to the other within a channel requires less activation energy than it does for a hydrated ion
to pass. Since these ions in aqueous solutions are strongly
hydrated, any such binding site must compete with the solvent
molecules. Accordingly, the affinity of binding is determined
by two opposing forces, the electrostatic attraction to the site
and the force of hydration. In other words, it is the difference
between the work gained through the binding to the site and the
work expended for the dehydration.

As to the detailed mechanism of selection the present views,
both based on the foregoing assumptions, go back to hypothesis
of EISENMAN (1961) and MULLINS (1959), respectively. Both these
hypotheses agree in that negative sites should attract dehydrated
Na^+ more strongly than they do dehydrated K^+ because the former
has the smaller size and hence the greater field strength of the
two. For the same reason, however, Na^+ binds the hydration water
more firmly than does K^+. In their explanations as to why some
sites prefer Na^+ and others K^+, the two views differ: Eisenman's
theory attributes the different preferences primarily to the
magnitude of the *field strength* of the binding site. If this is
relatively strong, the preference is primarily determined by the
site, the hydration being less important: Na^+ will be preferred
before K^+. On the other hand, if the field strength is relatively
weak, weaker than the force of hydration, the preference is determined by the dehydration, i.e., by the readiness of the ion
to part with its hydration water: Clearly, K^+ will be preferred
before Na^+. With binding sites of intermediate field strength,
different ions may be affected differently so that the order of
selectivity may change accordingly. For the five alkali ions,
Li, Na, K, Rb, Cs, 120 permutations in their sequence are mathematically possible. In biological systems, about 11 of these
have been verified and accounted for entirely in terms of field
strength and hydration work, in line with Eisenman's theory
(DIAMOND and WRIGHT, Annual Reviews of Physiology, 1969). The
fundamental validity of Eisenman's theory has gained support by
the construction of ion-specific glass electrodes, in which the
glass material is being composed on the basis of the above theory
(EISENMAN, 1975).

By contrast, in Mullin's view, as summarized more recently (1975)
the preference is largely determined by the *geometrical fit* of the
ion into the site. Between two ions with equal binding affinity
for a given site, the one with the better steric "fit" is postulated to be preferred. On this basis, the differential preference
with respect to K^+ and Na^+ is explained as follows: Let us assume
a channel of the shape of a pore just wide enough to be permeable
for a naked, i.e., dehydrated K ion. In the lining of this pore

there are polarizable groups whose affinity for the K ions is
high enough to compete with the hydration water, so that the K
ion may penetrate this channel. On the other hand, a Na ion,
even though it is smaller than the K ion in its dehydrated form,
may still have greater difficulty in entering and penetrating
this channel, because the affinity of the lining is not high
enough to dehydrate the Na ion, which owing to its small size,
does not "fit", i.e., it cannot come close enough to the binding
group, especially if the pores are rigid. So this channel clearly
should prefer K^+ to Na^+. Another channel is thinkable that is
narrower but just wide enough to permit a Na ion to enter and
that is fitted with groups strong enough to replace the hydration
shell of Na^+. Such a channel would transmit Na^+, but not K^+,
simply because K^+ is too big.

We see that in Mullin's theory selectivity involves competition
for the ion between hydration water and channel size, but depends
heavily on the geometry and the rigity of the channels or of the
steric structure of carrier sites.

3.1.3 *Translocation by Carriers.*

Most of the hypotheses offered to
explain the process of translocation are based either on *carrier*
models or on *channel* models. Available evidence seems to favor
the first alternative in some systems, and the second alternative
in others. Hence at present it would appear that both alternatives
exist side by side. For each alternative, experimental models are
available with artificial membranes and artificial translocators
(ionophores) as will be discussed elsewhere.

The carrier-model postulates a mobile component of the membrane
which is capable of reversibly binding the particle to be transported, and of oscillating between the two faces of the osmotic
barrier. It is usually depicted as a molecule that may freely
move within the membrane without leaving it, binding specifically
the transported solute species at one side and, after having
carried it through the barrier, releasing it at the other side
(Fig. 3a). The picture of a ferryboat is often used and found
to be useful as a basis for deriving kinetic equations of this
kind of transport. In reality, however, the mechanism of translocation may be quite different, especially if the carrier is a
protein molecule whose size exceeds the thickness of the lipid
phase (30 - 50 Å) which it has to penetrate. In other words, the
ferryboat would be greater than the "river". Because of these
and other difficulties, several modifications of the old ferryboat model have been suggested: For instance, that not the whole
molecule but only a "loose" chain swings like a flap from one
side of the barrier to the other; or that the carrier revolves
around an axis within itself, thereby causing a translatory
movement of the binding site if it is located in the periphery
of the carrier[6] (Fig. 3b). The "gate" model of PATLAK (1957)
deserves special mention because of its possible actuality, in

---

[6]The revolving carrier model is, according to more recent findings very unlikely since the rotation rate of membrane proteins around an axis parallel to the
lipid layer plain has found to be far too slow (SINGER, 1977).

Carrier models

I. Symmetrical

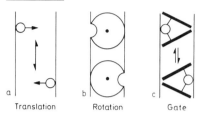

| a | b | c |
| Translation | Rotation | Gate |

II. Asymmetrical

d

Conformational

Fig. 3. Conventional carrier models. *Upper part* (a) A mobile carrier is cyclically translated as a whole across the membrane, thereby carrying the transported solute. (b) Carrier revolves around an axis inside itself, thereby moving the transport site with the transported solute between the two adjacent solutions. (c) Gate-type model of Patlack, each carrier consisting of two parts, performing tilting movements against each other, thereby being accessible to the transported solutes, alternatingly from the two adjacent solutions. (d) Conformational model. Due to a conformational rearrangement of tertiary protein structure the transporting binding site is moved between the two boundaries of the membrane. For further details see text

which a transport unit, by a tilting movement of its components, changes between two states, each of which makes a centrally located binding site accessible from a different side (Fig. 3c). This model is very similar to more recent ones, in which translocation of a binding site is attributed to a conformational change of the tertiary or quatary structure of a carrier protein (Fig. 3d). This might involve the rearrangement of peptide chains, so that the site-carrying part of a chain is translocated from a location close to one side to one close to the other side of the protein. We could accordingly speak of two conformational states of the molecule, an "extroverted" one in which the site-carrying chain is exposed to the one side, and an "introverted" one in which this chain is exposed to the other side of the barrier. An argument frequently raised against this carrier model is that a site translocation in space due to a conformational rearrangement of peptide chains could hardly exceed a distance of 10 Å, whereas the lipid layer is about 30 – 40 Å thick. This argument may not be cogent since at the places where a carrier protein is embedded in the lipid layer the critical barrier may be much smaller than the average thickness of the lipid layer. It is also possible that two carrier systems are arranged in series (LIEB and STEIN, 1972; LeFEVRE, 1973).

In the so-called tetrameric model of LIEB and STEIN (1972) two pairs of revolving carriers are arranged in series, with an empty space in between, into which the substrate has to be liberated before it can be captured by the opposing carrier. Of the pair of parallel carriers, one has a high affinity for the substrate

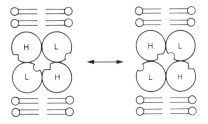

Fig. 4. Tetrameric model. Two pairs of hemicarriers H and L, with high affinity and low affinity for the transported substrate, respectively. The
tetramere can exist in either of two energetically equivalent conformations.
Transport occurs with the following sequence of events: (1) a sugar molecule
binds to one of the subunits, (2) following a conformational change, the
sugar will be present within the internal cavity of the tetramere, (3) the
sugar will distribute itself between the two inwardly facing binding sites
according to energetic considerations, (4) a second conformational change
occurs if, while in the internal cavity, the substrate had last been bound
to the binding site other than that on which it entered the cavity, then a
transport event will take place

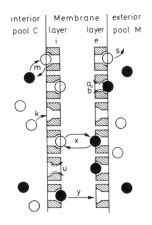

Fig. 5. Substrate-conditioned introversion model. Equal hemispheres are arranged in two layers, an internal one (i) and an external one (e). Each hemicarrier can independently switch between two positions, an intraverted one
and an extraverted one. The *hemiport*-like sites indiscriminately interact
with all suitable substrates (two species here indicated as *black* and *white*
balls) from the pool, either by simple association (k) or dissociation (s)
or by displacement of a substrate by another species. After a hemicarrier
with a substrate has switched to the introverted position, the substrate can
move to the semicarrier on the opposite side either by mere migration (y) or
by exchange (x) across the inner cavity

(H) and the other low affinity (L). Of the opposite pair, there
is also one high and one low affinity carrier, but the arrangement is such that the high affinity carrier of the outside pair
is in phase with the low affinity part of the inside pair, and

vice versa (Fig. 4). A special feature of this system is that
the four carriers arranged in that way do not revolve independently of each other, but synchronously. This model rests heavily
on many assumptions that cannot yet be tested experimentally.

In the model of LeFEVRE (1973), contrary to the preceding one,
all carriers have the same affinity for the solute, and the revolving movement of each is not linked to that of any neighboring
or opposite carrier (Fig. 5).

Kinetically, all models in their simple forms are hardly distinguishable, as long as they are symmetric. Hence each can be used
as a basis for the derivation of fundamental kinetic equations.
Such equations have been useful in the past to describe at least
qualitatively the behavior of many real transport systems under
various conditions. A different behavior, however, is to be expected if the system is asymmetric, if, for instance, the binding
affinity for the solute differs between the two sides.

The simplifying assumption of symmetry is not warranted in any
of the above models, but may be reasonable in those models in
which the carrier migrates or rotates as a whole. It is, however,
hardly justified in models postulating a conformational change.
The different conformational states of a protein are very unlikely to have the same properties, e.g., the same free energy
and the same affinity for the solute to be transported. Hence
such models are most likely asymmetric.

As will be shown later, the kinetics of asymmetric transport
systems are in many respects different from those of symmetric
ones, so that a distinction between these two alternatives might
be possible. Available evidence appears indeed to favor asymmetry for some systems or facilitated diffusion. Thermodynamically,
on the other hand, asymmetry should not introduce any difference.
An asymmetric system can therefore not perform active transport,
even if the affinities of the two conformational states for the
transported solute are different. According to the law of microscopic reversibility any such difference in affinity has to be
compensated for by appropriate differences in mobility of the
loaded and unloaded carrier in the forward and backward direction,
respectively, so that no active accumulation can occur.

3.1.4 Translocation Through Channels. In the last years, an alternative
kind of mediation has been considered that involves neither a
carrier nor a pore in the sense previously discussed. The mediator may be considered something in between carrier and pore in
that it combines some characteristic features of both. It has
been suggested to call this kind of mediator a "channel". Even
though these channels are envisaged as very small holes, there
is an increasing tendency to consider them something different
from pores in the conventional sense, even though there may be
transitional formations that could be classified as either pores
or channels. The main criteria of a channel in this sense as different from those of a common pore are derived mainly from observations with channel-like ionophores such as gramicidin in
artificial membranes, but there is evidence that analagous formations also occur in biological membranes:

Common pores are nonspecific, except that they become less permeable with increasing size of the penetrating particle. Channels,
by contrast, show some specificity that is not simply related
to the size of the passing solute.

Pores exert frictional resistance to the passing solute, which
should be the greater the smaller the pore size is compared to
the size of the passing particles. Hence the particle diffuses
through a pore more slowly than it would through an equivalent
volume element in free solution. A channel, by contrast, allows
very rapid penetration of the selected particle, even though the
width of the channel may not appreciably exceed the particle
size. Hence the particle may pass a channel much faster than it
would diffuse through an equivalent volume element in free solution. For example one specific channel of the squid axon nerve
membrane has a conductance of 10^{-10} mmhos, Ω^{-1}, corresponding to
a turnover of $\sim 10^{+8}$ ions per second at an electric PD of 100 mV.
It would require a water-filled pore in an equally thick membrane,
of more than 10 Å in diamter to allow the same turnover rate by
free diffusion in the same field (EHRENSTEIN and LECAR, 1972).
From all we know, the effective diameter of a specific channel
is probably smaller than 10 Å. Hence the solutes appears to pass
through their specific channel by a mechanism different from
free diffusion through an aqueous pore.

Pores are usually considered permanent features of the membrane
structure. Channels, on the other hand, are highly transient and
are usually attributed to special ionophoric substances that appear to oscillate rapidyl between a transmitting (open) and a
nontransmitting (closed) state, as between two conformational
states. Channels are therefore likely to be subject to regulatory
effects, be it by the transported solutes themselves, or by other
regulatory agents like hormones etc., electric PDs, etc. Such
effects may favor or disfavor the open, transmitting state over
the closed state.

In summary, channels are not merely holes but special devices
to promote specific and very rapid transfer.

It has been mentioned that the specificity of these channels is
attributed to the binding of the dehydrated ion to a specific
site. Whether the high rate of penetration is also due to such
a binding is not clear. If it is, the binding involved in the
penetration may not be the same as the binding necessary for the
specificity. There is evidence that specificity of a given channel and rate of penetration of the solute concerned are not related and sometimes even appear to be inversely proportional:
In other words, the higher the specificity, the slower is the
rate of penetration. It has been suggested that specificity is
strongest if the dehydration of the ion before binding is complete. The implication is that complete dehydration of an ion
has a high energy of activation and therefore delays penetration.
On the other hand, it is imaginable that the channel lumen contains an array of special binding sites along with the particle,
once it has entered the channel, may slide at high speed. It
thus seems that there are two kinds of sites involved, one accounting for the specificity and one for the rapid transmission.

In biological channels it is believed that the two kinds of sites
are geometrically separated, the specificity sites being located
all the entrance of a channel. Selective channels may have additional features like gates, i.e., transient constrictions that
may open or close according to circumstances, e.g., under the
influence of direction and magnitide of transmembrane potential,
chemical agents, etc. Some channels may even be temporarily
plugged up by certain ions, sometimes only from one side of the
membrane, as if there were a funnel-like opening able to bind
and fix an ion that is too big to penetrate the narrower part
of a channel.

It is assumed that the channel may change to the open state as
a "response" to the interaction with its specific ion. In other
words, the channel may become permeable to an ion only after this
ion has already been bound to a specific site. So a channel is
not just a rigid hole but a very reactive entity.

At this stage, however, to little is really known about the detailed structure of biological channels and their gates to justi-
fy a detailed discussion here.

Obviously, channels appear to have many properties in common
with mobile carriers. For instance, both accelerate penetration,
both are specific, and both are subject to competitive and non-
competitive inhibition. In both cases special substances, presumably proteins, are involved with which the solute has to interact in order to be transported. Whereas channels are superior
to carriers in accelerating translocation, the latter may have
the higher specificity.

A distinction between carriers and channels in biological membranes is difficult. A high turnover number per site and a low
temperature coefficient are usually taken to argue against a
mobile carrier. The extremely high rate at which certain membranes, such as nerve membranes, are penetrated by alkali ions,
is considered incompatible with the carrier mechanism and therefore calls for a channel mechanism.

On the other hand, many transport systems show phenomena that
fit a mobile carrier more easily than a channel. This, for instance, applies to the phenomenon of trans stimulation if it is
interpreted in terms of countertransport. To explain countertransport in terms of a channel would require some far-fetched
or unlikely assumptions. Also, many coupling phenomena, to be
discussed later, such as cotransport, seem to fit a carrier
mechanism more smoothly than a channel mechanism. The available
evidence, therefore, does not unequivocally prove the one and
disprove the other mechanism. It seems rather that both kinds
of mechanisms occur in biological membranes so that some transport systems function by a carrier and others, by a channel
mechanism. This apparent dualism is supported by the experiments
with ionophores in artificial lipid membranes. Both kinds of
ionophores appear to have been identified, those of the valino-
mycin type, which act as translocational carriers, and those of
the gramicidin type, which act by the formation of transient
channels. It could also be that there is only one type of mech-

anism present in biological membranes, which combines the characteristics of both channel and carrier, a kind of a "mobile channel" sometimes displaying mainly carrier-like and sometimes mainly channel-like behavior.

It is noteworthy that artificial ionophores of channel-like functions have so far been demonstrated only for ion movements. Whether this means that the channel mechanism is restricted to the ion movement or whether its possible function with nonionic molecules could not be demonstrated yet for technical reasons, cannot be decided at the present time.

The recent research on the subject has revealed many more details, which go beyond the scope of the present booklet. Many of the phenomena in this area are still awaiting their final explanation. The more interested reader is referred to the pertinent literature.

3.2   Treatment of Mediated Diffusion in Terms of the Law of Mass Action (LMA)

3.2.1 General. The kinetics of carrier transport, as used nowadays, is heavily indebted to the early pioneer work of WILBRAND and ROSENBERG, as summarized in 1961. The following treatment of carrier-mediated transport in terms of the LMA is based on their work, though their terms and symbols have been adapted to those preferred in the present booklet.

The rate equation for carrier-mediated transport are derived for a simple model system. As previously, we assume two well-stirred compartments separated by a membrane, each filled with an aqueous solution of the same solute, A, but at different concentrations on both sides of the membrane. Also here it is assumed that no net movement of water can take place so that the volumes of the two compartments remain constant, no matter how much solute passes the membrane. The new feature of this model is that the solute A cannot pass the membrane freely but only in combination with a mobile component of the membrane, generally called carrier. The carrier, X, is treated as if it were freely mobile within the membrane phase, but unable to leave it. In the interfaces X can react with the solute A of the solution to form the complex AX, which also can penetrate the membrane phase. If the concentrations of A in the adjacent solutions, a' and a", are different and below the saturation value, the activities of the complex (AX) in the two faces are also different, so that A will continuously be carried from the side of the higher to that of the lower concentration.

To describe the saturation phenomenon in a chemical process (S → P) one uses the concept developed by HENRI and by MICHAELIS for the saturation of enzymatic reactions:

The initial rate of such a reaction

$$v = V \frac{S}{S + K_m} \qquad\qquad (3.1)$$

K_m, usually called Michaelis constant, was originally thought to represent the dissociation constant of the enzyme substrate complex, ES; v is the initial reaction rate (p = o), and V the maximum value of this rate. If we apply this equation to the initial flow of a transportable substrate A through a membrane from one compartment into another we should obtain

$$J_a = J_a^{max} \frac{a'}{K_m + a'} \tag{3.2}$$

J_a denotes the flux rate, a' is the concentration of solute A in the cis compartment — the concentration of A on the trans side being zero (a" = o), K_m is called the "apparent Michaelis constant". A transport process differs from a simple enzyme reaction in that two translocation steps, that of the loaded and that of the unloaed carrier, respectively, are always involved. For instance, a transport carrier is not immediately available again after it has translocated its substrate: It has to return to the other interface of the membrane before accepting a new substrate. This applies also to the initial rate, i.e., when a" is negligible.

The transport process is more analogous to a reaction in which the enzyme undergoes a conformational or other change during each reaction, from which it has to recover before it can take up a new substrate. The rate of this "return trip" of the empty carrier, which may differ from the forward trip of the loaded carrier, has to be considered in all models involving a mobile carrier accounted for in the kinetic equation. Even for the initial rate (a" = o) the relative rate of the return trip of the empty carrier is implicit in the parameters J^{max} and K_m of the initial rate equation, which therefore have here a meaning somewhat different from the corresponding ones in enzyme reactions, as we shall see later.

To derive the complete transport equations, which describe the transport rate, J_a, as a function of the activities of the transported solute A in the bulk solutions, a' and a", of the total number of carrier sites, x_T, and of the "permeabilities" of the free and loaded carrier species, P_x and P_{ax}, respectively, we introduce another simplification into our model: The binding and dissociation reactions between carrier and transported solute in the interfaces of the membrane are much faster than the translocation of the carrier species through the barrier. As a consequence we may assume that there is almost equilibrium (quasi-equilibrium) between carrier and ligands all the time and that the overall transport rate is limited by the translocation rates of loaded and unloaded carrier species. Hence we can set

$$ax' = \frac{a' \cdot x''}{K_{ax}} \; ; \quad ax'' = \frac{a'' \cdot x''}{K_{ax}} \tag{3.3}$$

Before proceeding with the derivation we may try to characterize the symbols used here in view of the likely mechanics of a real system. At face value these symbols appear to stand for the concentrations of the carrier species in the phase boundaries (x and ax) and of permeability coefficients of these species within

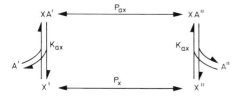

Fig. 6. Symmetric carrier system of facilitated diffusion (scheme). P_{ax} and P_x are mobilities (or probabilities of translocation) of loaded carrier (XA) and unloaded carrier (X), respectively. K_{ax} dissociation constant of carrier complex, assumed to be in *quasi-equilibrium* with free carrier and solutes in the adjacent solutions

the membrane phase (P_x, P_{ax}), respectively. Strictly speaking, this treatment is unrealistic, since it implies that within the membrane phase the concentration profiles are linear for each carrier species. These implications appear incompatible with what we know about the structure of the biological membrane, and also with more modern views on the probable nature and function of a carrier. Despite these inconsistancies, the kinetic equations derived on such an unrealistic basis are not only very popular but also appear to describe satisfactorily many observations with actual permeation processes. It looks as if the quantitative treatment were formally correct but as if the underlying concept of a carrier functioning as a ferryboat were not. We may therefore still use the above equations, but in order to put them on a more realistic basis, we should reinterpret the various symbols in terms of contemporary views. Accordingly, x', x", ax', ax" should represent the amount of carrier sites (in molar units per unit membrane) that are probable to exist in one of those four states. The permeability coefficients, P_x and P_{ax}, should in reality be transition probabilities, indicating the numbers of times that a single particle of the two carrier species, x and ax, is likely to be transferred from one interface to the other, or switches from the ' position to the " position and vice versa per time unit.

Also the rate coefficients, referring to the binding and releasing reaction between the carriers in the interface and the solutes dissolved in the adjacent solutions, can be regarded as probabilities of transition, i.e., to give the number of times that a binding or releasing process, respectively, is likely to occur per time unit and carrier particle. If we assume that all the transitions between the two interfaces are very rapid one-step transitions, the intermediate states of the carrier sites, i.e., on their way through the membrane, can be quantitatively neglected.

In order to be rate-limiting, the translocation steps have to be much slower than the interactions between carrier and solutes at the interface, which are taken to be in *quasi-equilibrium*. To reconcile the assumption of a comparatively slow translocation with the above-postulated very rapid one-step transitions we simply assume that the translocation steps occur very rarely, at least

much less frequently than do the interactions between carrier
in the interface and the dissolved solute A. Under these conditions a carrier site would remain long enough in each of the
four positions to allow quasi-equilibrium within the adjacent
solutions. Accordingly, the ratios of the coefficients for the
on- and off-reactions, respectively, in the interface can then
be treated like equilibrium constants (K_a' and K_a''). For convenience they are given the dimension of molar concentration.

The simplest procedure for deriving the kinetic equations is to
start from the following set of basic equations:

1. *The equation of conservation* is based on the assumption that the
total number of carrier sites that participate in the system
under consideration and is constant at all times (x_T). For reasons already discussed we only consider two positions for each
carrier species, that on the left side (cis) and that on the
right side (trans) of the barrier, indicated by ' and ", respectively. In other words, the number of carrier species in intermediate positions between the two interfaces of the membrane is
considered insignificant here with the implication that the tran
location at the molecular level is extremely fast but limits the
overall transport rate owing to the infequency of the transition

$$x' + x'' + ax' + ax'' = x_T.$$
(3.4)

2. *The steady-state equation* is based on the assumption that the total number of carrier species moving to the right side is identi
cal to the total number of carrier species moving in the opposit
direction during the same time interval. In the equation the sum
of all carrier species on the cis side, each multiplied by its
appropriate velocity (or probability) coefficient, is set equal
to the corresponding sum on the trans side.

$$P_{ax} \, ax' + P_x \, x' = P_{-ax} \, ax'' + P_{-x} \, x''$$
(3.5)

P_{ax} and P_x are the mobilities, or probabilities of transition,
of loaded and unloaded carrier, respectively, in the forward
(positive subscripts) and backward (negative subscripts) directions, respectively.

3. *The transport equations* give the rate of (coupled) translocation
of the carrier species containing the transported solute:

for the net transport

$$J_a = P_{ax} \, ax' - P_{-ax} \, ax''$$
(3.6)

and for the unidirectional fluxes

$$\overrightarrow{J}_a = P_{ax} \, ax'$$
(3.7a)

and

$$\overleftarrow{J}_a = P_{-ax} \, ax''$$
(3.7b)

From the preceding set of basic equations, the final equations
for the net movement and the unidirectional fluxes can be derive
by simple algebraic procedures, which need not be described here

Since these are sometimes tedious, simplifying short cuts are re-
commended such as those described by KING and ALTMAN (MAHLER and
CORDES, 1971).

To test a given model experimentally on the basis of these equations, biologists tend to start with the simplest model possible,
making as many simplifying assumptions as are plausible and aban-
doning these to the extent that they are not confirmed by experimental results. This procedure, though it may be less reliable
than the reverse one, which starts from the "complete" equation
and proceeds by introducing one simplification after another,
has often proved helpful to the intuitive understanding. We shall
therefore apply the "biological" approach here, and its various
stages happen also to be the historical ones.

3.2.2 Symmetric Carrier Systems . To start with the most simple carrier model, we make the following assumptions:

1. The rate coefficients (or probabilities of transition) are
the same for both the loaded and unloaed carrier ($P_{ax} = P_x$),

2. The system is symmetric, i.e., the corresponding equilibrium
constant and the rate coefficients of each carrier species are
the same on both sides of the membrane ($P_{ax} = P_{-ax}$, $P_x = P_{-x}$,
$K_a' = K_a''$).

Such a system can be described by two opposing Michaelis-Menten
equations with identical values for K_m and J_{max}. The total carrier species is always equally distributed between both sides
of the membrane

$$
\begin{aligned}
J_a &= P_x \frac{x_T}{2} \left(\frac{a'}{K_a + a'} - \frac{a''}{K_a + a''} \right) \\
&= P_x \frac{x_T}{2} \frac{K_a(a'-a'')}{K_a^2 + K_a(a'+a'') + a'a''}
\end{aligned}
\tag{3.8}
$$

After setting $P_x \frac{x_T}{2} = J_a^{max}$ we see that both initial rate J_a^o at
a" = o and unidirectional tracer flux $\vec{J}_a$ (at any a") in the forward direction are the same as in Eq. (3.2).

If we abandon one of the above simplifying assumptions, namely,
that of equal mobilities for the two carrier species, the carrier
is no longer equally distributed between the two sides of the
membrane: Its distribution depends on the difference in concentrations of substrate between the two sides of the membrane:

The *net flux* of this systems, as derived from Eq. (3.6) is

$$
J_a = \rho P_x x_T \frac{K_a(a' - a'')}{(K_a + a')(K_a + \rho a'') + (K_a + \rho a')(K_a + a'')}
\tag{3.9}
$$

$$
\rho = P_{ax}/P_x
$$

By setting a" = o we obtain the initial net flux in the forward direction:

$$J_a^o = \frac{\rho}{\rho+1} \; P_x \, x_T \; \frac{a'}{\frac{2}{\rho+1} K_a + a'}$$

with the standard parameters

$$J_a^{max} \quad \frac{\rho}{\rho+1} \, P_x \quad \text{and} \quad K_m = \frac{2}{\rho+1} K_a$$

The *unidirectional flux* in the same direction, however, is

$$\vec{J}_a = \rho P_x \, x_T \; \frac{(K_a + \rho a'') \, a'}{(K_a + a')(K_a + \rho a'') + (K_a + \rho a')(K_a + a'')} \tag{3.10}$$

with the corresponding standard parameters:

$$\vec{J}_a^{max} = \frac{\rho(K_a + \rho a'')}{(\rho+1) K_a + 2\rho a''} \, P x_T \tag{3.11}$$

$$\vec{K}_m = - \, K_a \; \frac{2K_a + (\rho+1) a''}{(\rho+1) K_a + 2\rho a''} \tag{3.12}$$

The standard parameters of the unidirectional flux differ from those of the corresponding net flux of the same system in that they depend on a", the "trans"concentration of A. These differences vanish if a" = 0, or if ρ = 1 (i.e., if $P_{ax} = P_x$); the parameters of the unidirectional flux then become identical with those of the corresponding net flux.

Since the system is symmetric, the parameters of the initial net flux and of the unidirectional flux, respectively, are numerically identical in the forward and backward direction, except that a' and a" have to be properly interchanged.

It is seen that the unidirectional fluxes differ from the corresponding initial net fluxes, except at ρ = 1, i.e., at equal mobilities of loaded and unloaded carrier species, and except at zero trans concentration. Otherwise each unidirectional flux depends on both the cis and trans concentrations of A. If ρ > 1, the flux is stimulated by the trans concentration (trans stimulation), whereas the opposite is true if ρ < 1 (trans inhibition). Hence, this model accounts for the trans effects observed with many transport systems (HEINZ, 1954; HEINZ and DURBIN, 1957; ROSENBERG and WILBRANDT, 1957; HEINZ and WALSH, 1958).

The flux ratio

$$\boldsymbol{f}_a = \frac{\left(\frac{K_a}{\rho} + a''\right) a'}{\left(\frac{K_a}{\rho} + a'\right) a''} \tag{3.13}$$

is clearly smaller than $\frac{a'}{a''}$ if a' > a", as is typical for negative tracer coupling, which will be discussed later. It becomes unity if a' = a". The flux ratio is also different from the concentration ratio if $\rho = 1$, while at higher ρ-values the flux ratio converges towards unity. The deviation of the flux ratio from the concentration ratio is primarily due to the saturability of the system.

3.2.3 Asymmetric Carrier Systems.

The kinetic equations become still more involved if we abandon another simplifying assumption, namely, that of symmetry. Systems are likely to be asymmetric to the extent that the translocation of solute is brought about by a conformational change rather than by a translation or rotation of the carrier protein as a whole, since it is highly improbable that two conformational states of a protein have equal parameters. For instance, the affinity of the binding site for the substrate (K_a) is likely to change during the transition from one conformation to another. This difference in K_a between the different sides of the membrane must, however, not lead to change in equilibrium distribution of the substrate between the two sides of the membrane, which would mean uphill transport without the expenditure of energy. Accordingly, the various rate constants will adjust to each other so that the law of detailed balance is not violated, as will be shown below. The kinetics of asymmetric systems have been extensively and rigorously studied by GECK (1971). The present treatment is based on these studies but has been greatly simplified and translated into the notation preferred in this book.

In the asymmetric model, in contrast to a symmetric model, the unidirectional fluxes are no longer identical in the forward and backward direction for each carrier species, nor are the apparent Michaelis constants identical for the two sides of the barrier. The asymmetric model in its simplest form may be represented by the diagram shown in Figure 7.

The double arrows are to indicate that the unidirectional rate constants may be different for each step. P_{ax}, P_x denote translocation coefficients, k_1 and k_2, reaction rate coefficients. Positive subscripts refer to the forward (or upward) direction, negative subscripts to the backward (or downward) direction. The differences between the forward and backward rate coefficients of the various steps may vary a great deal, but are restrained by the principle of detailed balance. The second law of thermodynamics requires that the final ratio of product to substrate must not exceed the equilibrium constant. In addition, the prin-

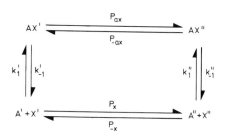

Fig. 7. Asymmetric carrier system for facilitated diffusion. P_{ax} and P_x as before, the negative subscripts indicating negative (*right to left*) direction of the translocation. k_1 and k_{-1} rate coefficients or association (on) and dissociation (off) of the carrier-substrate complex

ciple of detailed balance requires that the product of all rate coefficients in the clockwise direction, divided by the product of the all rate coefficients in the counterclockwise direction, be equal to the equilibrium constant. Since for facilitated diffusion of a neutral solute equilibrium distribution is unity, it follows that the two products of the rate coefficients are equal:

$$P_{ax} \cdot k''_{-1} \cdot P_{-x} \cdot k'_1 = P_{-ax} \cdot k'_{-1} \cdot P_x \cdot k''_1 \tag{3.14}$$

As in the figure, the positive subscripts denote the forward, the negative subscripts, the backward flow. ' and " refer to the left and right sides, respectively, of the system. For simplification we may again assume that the translocation occurs much less frequently than the interaction between carrier and solute in the interfaces, so that we can expect quasi-equilibrium of the latter reactions. The equation of detailed balance simplifies then to

$$P_{ax} \cdot P_{-x} \cdot K''_{ax} = P_{-ax} \cdot P_x \cdot K'_{ax} = \pi \tag{3.15}$$

equilibrium constants K'_a and K''_a replacing the ratios of the coefficients of the on- and off-reactions, respectively. π stands for the product at detailed balance. The final equations of facilitated diffusion in such a system are, for the net transport

$$J_a^{net} = P_{ax} \; X_T \; \frac{\pi \; (a'-a'')}{\Sigma} \tag{3.16}$$

The initial net fluxes (zero trans) for the forward direction ($a'' = 0$)

$$_0\overrightarrow{J}_a = \frac{P_{ax} P_{-x}}{P_{ax}+P_{-x}} \; X_T \; \frac{a'}{a' + \dfrac{P_x+P_{-x}}{P_{ax}+P_{-x}} \; K'_{ax}} \tag{3.17a}$$

and for the backward direction ($a' = 0$)

$$_0\overleftarrow{J}_a = - \frac{P_{-ax} P_x}{P_{-ax}+P_x} \; X_T \; \frac{a''}{a'' + \dfrac{P_x+P_{-x}}{P_{-ax}+P_x} \; K''_{ax}} \tag{3.17b}$$

are asymmetric.

The unidirectional fluxes are clearly different from the corresponding initial net fluxes as was the case with the symmetric system with unequal mobilities of loaded and unloaded carriers [Eq. (3.10a,b)].

$$\overrightarrow{J}_a = P_{ax} \; X_T \; \frac{(P_{-x}K''_{ax} + P_{-ax} a'') \; a'}{\Sigma} \quad \text{and} \tag{3.18a}$$

$$\overleftarrow{J}_a = P_{-ax} \; X_T \; \frac{(P_x K'_{ax} + P_{ax} a') \; a''}{\Sigma} \tag{3.18b}$$

$$\Sigma \;=\; (P_x + P_{-x})\, K'_{ax} \cdot K''_{ax} + (P_{ax} + P_{-x})\, K''_{ax}\, a' + (P_{-ax} + P_x)\, K'_{ax}\, a'' +$$

$$(P_{ax} + P_{-ax})\, a'\, a''$$

Even at equal cis and trans concentrations of A, the two *net fluxes* are not identical, in contrast to the corresponding fluxes in symmetric systems. As a consequence, such a system may have rectifying properties in that the net flow of A in the one direction from that in the opposite direction, even if the concentration differs from that in the opposite direction, even if the concentration differences are numerically equal. The asymmetry is most salient with the "standard parameters", i.e., maximum net flux ($o\overrightarrow{J}_a^{max}$) and apparent Michaelis constant (K_m) at zero trans concentration (z.t.), as shown below:

$$o\overrightarrow{J}_a^{max} \;=\; \frac{P_{ax} \cdot P_{-x}}{P_{ax} + P_{-x}}\, x_T \; ; \qquad o\overleftarrow{J}_a^{max} \;=\; \frac{P_{-ax} \cdot P_x}{P_{-ax} + P_x}\, x_T \qquad (3.19)$$

$$K'_m \;=\; \frac{P_x + P_{-x}}{P_{ax} + P_{-x}}\, K'_{ax} \; ; \qquad K''_m \;=\; \frac{P_x + P_{-x}}{P_{-ax} + P_x}\, K''_{ax}$$

It is seen that for each of these parameters different values are obtained for different directions. By inserting proper coefficients, the rectification can be increased almost indefinitely, without violating the law of microscopic reversibility. This asymmetry declines as the system approaches equilibrium, having its maximum value with the maximum net flow at zero trans concentration.

The asymmetry does not show in the *flux ratio*: In other words, the same ratio of the concentrations of A in the two compartments will give the same flux ratio since

$$\boldsymbol{f}_a \;=\; \frac{\left(\dfrac{K''_{ax} \cdot P_{-x}}{P_{-ax}} + a'' \right) a'}{\left(\dfrac{K'_{ax} \cdot P_x}{P_{ax}} + a' \right) a''} \qquad (3.21)$$

where, according to the law of detailed balance Eq. (3.15)

$$\frac{K''_{ax}\, P_{-x}}{P_{-ax}} \;=\; \frac{K'_{ax}\, P_x}{P_{ax}}$$

Consequently, the asymmetry does not show in the equilibrium exchange, i.e., the tracer fluxes at equilibrium of A (a' = a") are equal in either direction.

In an attempt to introduce a quantitative measure for the degree of asymmetry, the ratio Q of the two maximum net flows in the forward and backward direction, respectively, has been introduced (LIEB and STEIN, 1972). Accordingly we get the following relationship:

$$Q = \left(\frac{\vec{J}_a^{max}}{-\overset{\leftarrow}{J}_a^{max}}\right)_{z.t.} = \frac{K'_{ax}(P_{-ax}+P_x)}{K''_{ax}(P_{ax}+P_{-x})} = \left(\frac{K'_m}{K''_m}\right)_{z.t.} \tag{3.22}$$

Since the asymmetry effects vanish as a system comes close to equilibrium, they should have only a minor effect in the representation of these flows in terms of irreversible thermodynamics, provided that the system is close enough to thermodynamic equilibrium. To the extent, however, that irreversible thermodynamics is applied beyond this range, as can be done under certain conditions, we should expect asymmetry effects in the linear phenomenological coefficients (L), as will be shown below.

3.2.4 A Special System. To test the carrier model experimentally, the standard parameters, i.e., the maximal transport rate (J_{max}) and the half saturation (Michaelis) constant (K_m), are determined under various conditions and fitted with the corresponding parameters derived from the kinetic equations for the model concerned. The facilitated diffusion of D-glucose in human red blood cells, probably the most extensively studied system of this kind, may serve as a good example for this procedure. The most common methods for determining these parameters experimentally, which have been extensively applied to the facilitated diffusion of D-glucose in human red blood cells, include:

1. The net exit of glucose from cells excessively loaded with glucose into a medium with varying glucose concentration. The exit at zero concentration in the medium would give the maximum rate (J_{max}) and the concentration necessary to reduce this rate to half this value is equated with K_m. This method has been introduced by SEN and WIDDAS (1962) and is also called *infinite cis* (i.c.). It requires that the intracellular glucose concentration at the time of measurement is sufficiently above that given by the Michaelis constant, so that maximum net exit in the absence of extracellular sugar is warranted.

2. The net outflow into a medium free of glucose from cells containing glucose at various concentrations. The intracellular concentration at which this exit has half its maximal value would be K_m. This method is also called *zero trans* (z.t.)

3. Both parameters can also be determined for the (unidirectional) tracer fluxes of glucose into cells pre-equilibrated with sugar at various concentrations — equilibrium exchange (e.e.)

4. K_m should also be equal to the concentration of extracellular glucose required for half maximum inhibition of influx of a low affinity substrate of the same system, e.g., sorbose.

It came first as a great surprise that the parameters for the above system, if determined by the different methods, showed persistent differences, which could be reproduced by many laboratories at different times (MILLER, 1968a,b). Most striking is the difference between the K_m determined by methods (i.c.) and 2 (z.t.), respectively, method 2 always giving values more than ten times greater than those obtained by method 1 (Table 3). Any model that does not account for these discrepancies should be

Table 3. Parameters of D-glucose transport in human and red blood cells

Method	K_m (mM)	J_{max} (mmol/min/cell unit)
"Sen - Widdas" (infinite cell)	1.8^a	140^a
Net exit (zero trans)	25.0^a	
Equilibrium exchange	38.0^a	260^a
Sorbose inhibition	23.0^b	

^{a}LIEB and STEIN, 1972 ^{b}MILLER, 1971

modified or rejected. Let us now test the described versions of the carrier model in this respect.

The simplest model [Eq. (3.8)] predicts that all methods should give identical values for J_{max} and K_m, and can therefore be rejected. The next model, which differs from the simplest one only by different rate coefficients for loaded and unloaded carrier ($P_{ax} > P_x$), shows equal J_{max} by methods 1 (i.c.) and 2 (z.t) but higher values by method 3 (e.e.). The K_m values obtained by method 2 (z.t.) should be slightly smaller than the K_m obtained by method 1 (i.c.), and that obtained by method 3 (e.e.) should exceed the latter but by not more than a factor of two. Neither the extent nor the direction of these deviations of K_m is compatible with those found experimentally: K_m as determined by method 2 however is ten times *larger* than K_m determined by method 1. Hence also this version has to be rejected, even though it explains some phenomena, such as accelerating trans stimulation, better than the preceding model.

One can proceed further and abandon also the assumption of quasi-equilibrium in the membrane faces by allowing that the association and dissociation reactions between carrier and solutes of the adjacent solutions may become rate limiting, rather than the translocation step, but in spite of some improvement, the major discrepancies between observation and kinetic equation cannot be fully removed by this extension either (BLUMENTHAL and KATCHALSKY, 1969).

Several attempts have been made to modify or extend the original carrier model in order to account for these discrepancies. None of these attempts has been satisfactory in all respects, but the principles underlying such modifications are nonetheless worth consideration. These are:

a) Unstirred layers outside the cell, which could to some extent account for the discrepancies observed. It could be shown, however, that owing to the much greater velocity of free diffusion across a thin unstirred layer as compared to the rate of facilitated diffusion across the membrane, true unstirred layer effects are only of minor importance here (MILLER, 1972). Still, the phenomenon cannot be discarded entirely, since any resistance in series with the transport — which is what matters here — might give rise to effects that are indistinguishable from those of unstirred layers. There could, for instance, be a barrier between the reacting transport site and the adjacent solution; or the substrate, before reacting with the carrier, might have to be dehydrated, a process that may have an activating energy high enough to impede the reaction with the carrier in a way that could be indistinguishable from an unstirred layer. Such a barrier could occur inside and outside the cell membrane.

b) Also the previously mentioned hemicarrier systems of LIEB and STEIN (1972) and of LeFEVRE (1973) have been considered in this context.

Apparently none of the above listed modifications alone can explain all facts, but certain combinations of them might do. So a membrane may at the same time be somewhat asymmetric and may possess in addition some features causing effects similar to those of an unstirred layer (EDWARDS, 1974). It would thus appear that the classical carrier model as such cannot be discarded yet merely on kinetic grounds, since not all possibilities for introducing combinations of modifying features have yet been exploited.

A more recent reappraisal of the problem by FOSTER and JACQUEZ (1976) concludes that asymmetry of the membrane parameters (MILLER, 1971) should account for the discrepancies observed.

3.2.5 Channel Systems. When considering the kinetics of channel-mediated transport, some preliminary information may be obtained from the kinetics of "single-file diffusion" (HECKMANN, 1964) under the reasonable assumption that single-file diffusion, wherever observed, is in reality channel-mediated transport.

The phenomenon of single-file diffusion requires that the particle, while passing the pore, interacts with at least two binding sites, and that two particles cannot exchange places, i.e., one particle can jump from site 1 to site 2 only if site 2 is free, or if another particle, occupying site 2, leaves this by jumping out of the membrane (Fig. 8). Heckmann's explicit equation for single file net flow, translated into the notation used in this

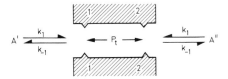

Fig. 8. Model of single-file diffusion (ungated channel). Two fixed binding sites, *1* and *2*. P_t rate constant (or probability) of transition of solute between the two binding sites. k_1 and k_{-1} rate constants for the association and dissociation reactions between solute *A* and the binding sites

book is:

$$J_a = P_t x_c \frac{K_a(a' - a'')}{(K_a + a')(K_a + a'') + \frac{P_t}{k_1}(2K_a + a' + a'')} \qquad (3.23)$$

P_t is the rate coefficient for the translocation step $(1 \to 2)$ and K_a is $\frac{k_{-1}}{k_1}$, the ratio of the rate coefficients of the binding (k_1) and the dissociation reaction between each site $(1$ and $2)$ and the solutes dissolved in the adjacent solutions. x_c is the number of channels per unit area.

It is identical with that of the simplest carrier model for the special case that the rate of the translocation step (P_t) is not negligible as compared to the rates of the binding and associa- tion reactions. It predicts saturability of the initial net flow with the maximum rate:

$$\overset{o}{J}_a^{max} = \frac{P_t x_c \cdot K_a}{K_a + \frac{P_t}{k_1}} \qquad (3.24)$$

and the apparent Michaelis constant:

$$K_m = K_a \frac{K_a + 2\frac{P_t}{k_1}}{K_a + \frac{P_t}{k_1}} \qquad (3.25$$

If, according to the assumption made for carrier transport, P_t is small as compared to k_1 and k_{-1}, $\overset{o}{J}_a$ reduces to simple Michaelis- Menten equation

$$\overset{o}{J}_a = P_t x_c \frac{a'}{K_a + a'}$$

In view of the extreme rapidity attributed to channel transport, however, one may speculate that here P_t may be much greater than k_1, so that we obtain for the net movement:

$$J_a = \frac{x_c}{2} \frac{k_{-1}(a' - a'')}{(2K_a + a' + a'')} \qquad (3.26)$$

The overall net rate is limited here by the speed at which A gets off the site at the trans side (k_{-1}). The maximum initial rate would accordingly be $\frac{k_{-1}x_c}{2}$ and the Michaelis constant, $2K_a$.

In contrast to the equation of net flow, the equations of the unidirectional fluxes in single-file diffusion are distinctly different from those in carrier mediated transport. From HECKMANN's (1964) equations we may derive the following equations for the true unidirectional tracer fluxes, expressed in our own symbols and simplified according to the condition that each tracer flux is determined at zero specific activity on the trans side:

$$\overrightarrow{J}_a = P_t x_c \frac{K_a}{\Sigma}\ a'\ (1- \frac{P_t}{k_1}\ \gamma a") \tag{3.27a}$$

$$\overleftarrow{J}_a = P_t x_c \frac{K_a}{\Sigma}\ a"\ (1- \frac{P_t}{k_1}\ \gamma a') \tag{3.27b}$$

$$\Sigma = (K_a + a')(K_a + a") + \frac{P_t}{k_1}\ (2K_a + a' + a") \tag{3.28}$$

$$\gamma = \frac{2(a' + a" + 4K_a)}{(a'+a"+2K_a)\left[2\frac{P_t}{k_1}(a'+a"+4K_a) + (a'+2K_a)(a"+2K_a)\right]}$$

Inserting this expression for γ we get very involved equations for the unidirectional fluxes, which can be analyzed to show transinhibition. For better understanding we present two border-line cases

1. $P_t \gg k_1$

$$\overrightarrow{J}_a = k_{-1}x_c\ \frac{a'(a'+2K_a)}{4(\bar{a}+K_a)^2} \tag{3.29}$$

Obviously, the flux is limited by the rate at which the label leaves the channel on the trans side (k_{-1})

2. $P_t \ll k_1$

$$\overrightarrow{J}_a = P_t x_c\ \frac{K_a\ a'}{(a'+K_a)(a"+K_a)} \tag{3.30}$$

Obviously, the flux is now limited by P_t, the rate at which the particle A jumps from one site to the other within the channel.

The flux ratio for the general case would be, setting $\frac{a'+a"}{2} = \bar{a}$,

$$f_a = \frac{a'(a'+2K_a)\left[a"+2K_a+2\frac{P_t}{k_1}\frac{\bar{a}+2K_a}{\bar{a}+K_a}\right]}{a"(a"+2K_a)\left[a'+2K_a+2\frac{P_t}{k_1}\frac{\bar{a}+2K_a}{\bar{a}+K_a}\right]} \tag{3.31}$$

According to expectation it shows positive coupling, as will be shown in Chapter 4.

A special situation is bound to occur if we remove one of the above restrictions by permitting, or even facilitating, the ex-change of place of solute particles between two neighboring sites That such "facilitated exchange" really occurs in biological channels seems unlikely though not a priori unreasonable. Under these conditions, trans stimulation should be expected. Such con-siderations are still highly speculative and can probably not be tested yet; thus, it might be premature to consider trans stimu-lation a proof of a mobile carrier, as has previously been done.

4 Isotope Interaction – Tracer Coupling

4.1 Mechanistic Aspects of Solute-Solute Interaction

4.1.1 General. In the previous treatment of solute flows, interactions of the passing solute particles with other substances, except for the transient interactions with fixed of mobile membrane constituents (solute-membrane interactions), have been disregarded. In biological membranes, however, a penetrating solute particle is likely to interact with other moving particles, either of the same, or of a different solute species. Such interactions may be direct, e.g., by friction and collisions between moving particles, or by chemical reactions between the particles to form new, more permeant species, or they may be indirect, i.e., mediated by a carrier or channel mechanism.

If the interaction is between distinguishable particles, i.e., between particles of different chemical species or between different isotopes of the same species, phenomena of coupling between the flows of the different species or isotopes, respectively, may be observed. Before dealing with the mechanism of such coupling we shall have a brief look at some typical coupling phenomena.

4.1.2 Cis and Trans Effects. Is has been known for a long time that the transport of a given solute across a biological membrane may be influenced, i.e., stimulated or inhibited, respectively, by the presence of another solute, which may be of a different species or an isotope of the same species. In some cases, the second solute has to be on the cis-side, i.e., the side for which the flow of the test solute originates, in order to be effective (cis stimulation or cis inhibition, respectively). In other cases, the second solute is effective from the trans side, i.e., from the side of the membrane toward which the flow of the test solute is directed (trans stimulation and trans inhibition, respectively). Sometimes the same second solute inhibits, if present on the cis side, and stimulates, if present on the trans side. It appears that cis inhibition and trans stimulation are the most frequent combinations observed, but also cis stimulation and trans inhibition, respectively, have been described.

Cis and trans effects are strongly suggestive of coupling between flows. It should be stressed, however, that the four terms mentioned above are strictly phenomenological: They merely describe what is observed with a given flux upon the addition of another solute to the cis or the trans solution. They are therefore not necessarily synonymous with other terms often used in this context, such as cotransport or symport for cis stimulation, countertransport or antiport for trans stimulation, and competition for cis inhibition. These other terms already imply interaction of flows and interpret the observed effects as coupling phenomena in terms of preconceived transport models.

Hence cis and trans effects are not yet conclusive evidence of the validity of the hypothetical mechanism. So the stimulatory and inhibitory effect could be due to structural alterations of the membrane permeability induced by the interacting solute on the appropriate side. In viable biological systems, also regulatory (feedback) effects of certain solutes have to be considered. Cases have been described in which the presence of a transported solute inside the cell may directly affect the transport of the same or another solute into the cell by some "feedback" mechanism (RING and HEINZ, 1966).

In summary, cis and trans effects are consistent with, and may even strongly suggest, coupling but are not sufficient to prove it. For this purpose the two flows supposedly coupled to each other should be studied directly, as will be discussed in Chapter 5.

4.1.3 Coupling Phenomena. The coupling between the flows of different solute species through biological membranes, which may lead to the active transport of a solute species, shall be treated in Chapter 6. In the present context we shall confine ourselves to a single solute species and discuss only isotope interaction. We speak of isotope interaction if the solute particles, while penetrating the membrane, interact with other particles of the same species, moving in the same or the opposite direction. Such interaction may influence the net flow of either solute species concerned, in particular it may disturb the proportionally between net flow and concentration difference, which is usually postulated for free diffusion. Cooperative as well as antagonisti effects may become apparent: The former may cause the rate to rise with increasing concentration gradient exponentially rather than linearly. The latter, by contrast, do the opposite: They may cause the rate to tend toward a maximum with increasing concentration gradient (saturation).

Isotope interaction is most conspicuous with the unidirectional fluxes as determined by isotopically labeled tracers, rather than with the net transport. Under these conditions typical coupling phenomena between the opposing tracer fluxes (tracer coupling) may be observed.

Owing to isotope interaction the unidirectional fluxes, as studied with tracers, should be subject to cis and trans effects and hence depend on the concentrations of the abundant isotope on the cis and trans sides, respectively, of the barrier. Accordingly, as has been shown with the equation of facilitated diffusion, the unidirectional fluxes may have parameters, K_m and J_{max}, different from those of the corresponding initial net fluxes [Eqs. (3.11, 3.12)]. Obviously the interaction between different isotopes of the same solute species may produce phenomena that are in various respects similar to those of true coupling, for instance coupling between the flows of different species. Hence the behavior of tracer fluxes may sometimes mislead one to postulate

energetic coupling, as in active transport, where there is none.
It is therefore important to differentiate between the coupling
phenomena of isotope interaction, also called "tracer coupling"
and those of true coupling. The similarity between these two
kinds of coupling is indeed limited to certain conditions so
that they can fundamentally be distinguished as will subsequently be shown.

Tracer coupling is preferentially studied with flux ratios rather
than with individual unidirectional fluxes, mainly because some
unknown parameters, to the extent that they equally affect fluxes
in both directions, cancel. In addition, flux ratios can be
treated in terms of TIP, which is not possible with unidirectional fluxes.

The basic procedure for studying tracer coupling is to relate
the flux ratio to the corresponding ratio of the electrochemical
activities of the solute under consideration. We recall that in
free diffusion both ratios should be equal. This is not so in
the presence of isotope interaction, except in true equilibrium,
i.e., if the electrochemical activities of the solute are the
same on both sides of the membrane. Hence any deviation of the
flux ratio from the corresponding activity ratio could indicate
isotope interaction or true coupling, or both. In view of this
similarity one would be justified in considering isotope interaction as a special kind of coupling, hence the term tracer coupling. On the other hand, as has been mentioned, there is one
important difference between tracer coupling and real coupling:
At zero driving force tracer coupling should disappear, but true
coupling should not.

In analogy to true coupling, we can distinguish between *positive*
and *negative* tracer coupling: We speak of positive coupling if
the driver flow tends to drive the coupled flow in the same di-

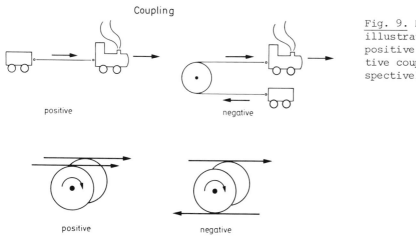

Coupling

positive   negative

positive   negative

Fig. 9. Mechanistic
illustration of
positive and nega-
tive coupling, re-
spectively

rection, and of negative coupling, if it does so in the opposite direction. This may be illustrated by a simple model (Fig. 9).

Accordingly, coupling via cotransport, for example, would be positive, and coupling via countertransport, negative. This distinction cannot simply be applied to unidirectional fluxes, which by definition have only *one* direction and cannot be inverted. Furthermore, to the extent that we distinguish only between two opposing unidirectional fluxes, the analog of cotransport does not exist here. The terms positive and negative tracer coupling are therefore used here only symptomatically; i.e., merely to indicate that the two opposing fluxes stimulate each other (negative coupling) or inhibit each other (positive coupling).

4.1.4 Effects on Flux Ratio. The sign of tracer coupling can be derived from the direction in which the flux ratio deviates from the activity ratio: If the latter is different from unity, the flux ratio will differ more from unity in positive coupling, and less so in negative coupling. In other words, if the activity ratio is greater than unity, it will be smaller than the flux ratio in positive tracer coupling, but greater than the flux ratio in negative coupling. If, however, the activity ratio is smaller than unity, it will be the other way round.

In mathematically more precise terms this would mean that in positive coupling the logarithm of the flux ratio would exceed numerically that of the activity ratio, whereas in negative coupling the converse would be true:

	Positive Coupling	*Negative Coupling*
	$\boldsymbol{f}_a > \dfrac{a'}{a''} > 1$	$\boldsymbol{f}_a < \dfrac{a'}{a''} > 1$
or	$\boldsymbol{f}_a < \dfrac{a'}{a''} < 1$	$\boldsymbol{f}_a > \dfrac{a'}{a''} < 1$
or	$\left| \log \boldsymbol{f}_a \right| > \left| \log \left(\dfrac{a''}{a'} \right) \right|$	$\left| \log \boldsymbol{f}_a \right| < \left| \log \left(\dfrac{a''}{a'} \right) \right|$

The interaction between penetrating particles of *different solute species* should follow the same principles as does isotope interaction. Also here the interaction may lead to stimulation or inhibition of the penetration, or even to uphill movement of the test species. The interaction by friction is here probably less significant biologically than that by a chemical reaction between the interacting particles leading to the formation of new, more permeant species such as neutral or lipophilic molecules or complexes from less permeant electrolyte ions. Interaction between different solute species', however, may, in contrast to isotope interaction, lead to true coupling phenomena with the possible transfer of energy, similar to those in secondary active transport, which will be dealt with more extensively in the subsequent chapter on Coupled Transport (Chap. 7).

4.2 Treatment of Tracer Coupling (Isotope Interaction) in Terms of the Law of Mass Action (LMA)

In the following, the flux ratio will be derived from the corresponding kinetic equations for some typical cases of interaction between penetrating solutes of the same species.

4.2.1 *Friction*:
As has been pointed out before, friction between freely moving particles in biological fluids is considered small owing to the high dilution of solutes in biological fluids. It may become significant, however, in membranes whose pores are very narrow relative to the effective size of the penetrating particles. With homoporous membranes that have uniform cylindrical pores of defined size it could be shown that interaction may occur between particles of sufficient size, like diffusing protein molecules. Whether similar friction occurs between the smaller solutes in the pores of biological membranes is still questionable, though some observations have been made that were termed single-file diffusion. Somewhat more realistic might be the interpretation that the particles in such a case do not simply diffuse through a water-filled pore but "jump" between more or less specific binding sites of a *channel*. The resulting tracer coupling should be positive, because the flux ratio tends to exceed the ratio of the solute activities, if the latter is greater than unity.

This is shown by the flux ratio for single-file (or channel) diffusion, derived from Heckmann's equation for the unidirectional fluxes $\left[$Eq. (3.31)$\right]$. f_a exceeds $\frac{a'}{a''}$ if a' > a". The difference vanishes if $(\frac{P_t}{k_1})$ becomes very small, i.e., if the rate at which the solute particle moves within the channel (P_t) is small compared to the rate at which it enters the channel (k_1). Apparently f_a cannot become smaller than $(\frac{a'}{a''})$ as long as a' > a"; in other words, coupling cannot become negative for this model, not even at saturation concentrations of a' and a".

4.2.2 *Association (Dimerization)*:
If the solute particles have a tendency to form dimers within the membrane and if these dimers pass the membrane more readily than do the monomers, positive isotope interaction should be expected. This is borne out by the flux ratio of dimer penetration.

$$f_a = \frac{\frac{2P_{aa}}{K_a}(a')^2 + P_a a'}{\frac{2P_{aa}}{K_a}(a'')^2 + P_a a''} = \frac{a'}{a''} \frac{(2P_{aa} \cdot a' + P_a K_a)}{(2P_{aa} \cdot a'' + P_a K_a)}$$

P_{aa} stands for the penetration rate coefficient of the dimer (a_2) and P_a for that of the monomer (a). K_a is the dissociation constant of the dimer: $(a^2/a_2) = K_a$. Also here f_a exceeds (a'/a") if a' > a", unless P_a is very large as compared to (P_{aa}/K_a), i.e., if the dimer were to contribute little to the overall penetration.

4.2.3 Carrier Mediation: The most likely kind of isotope interaction appears to occur in facilitated diffusion through the membrane. Saturation generally causes a deviation of the *flux ratio* from the activity ratio. However, independent of saturation, the flux ratio depends also on the difference in mobility between loaded and unloaded carrier. If the loaded carrier moves faster than the unloaded carrier, as seems to be true for many systems with accelerative counterflow, negative tracer coupling may appear, i.e., the flux ratio is closer to unity than the activity ratio. The opposite is to be expected if the unloaded carrier moves faster than the loaded one, a condition which should lead to transinhibition but seems to occur rarely in biological transport systems. If each carrier has only one binding site for a solute particle, these particles may inhibit each other's transport by competition for these sites, more so if the concentration of the solute rises toward saturation concentration. Accordingly the flux ratio will converge toward unity if the carrier becomes saturated on both sides of the membrane. The flux ratio, as derived from previous equations for facilitated diffusion, is

$$\mathit{f}_a = \frac{a' (K_a + \rho a'')}{a'' (K_a + \rho a')}$$

As previously, ρ stands for the ratio of the rate coefficient of the loaded carrier over that of the unloaded one.

$$\rho = \frac{P_{ax}}{P_x}$$

In contrast to the preceding examples, the flux ratio is here smaller than the activity ratio, if $a' > a''$, which is typical for negative tracer coupling. This applies also to asymmetric systems provided that ρ and K_a are appropriately chosen.

If ρ is greater than unity, coupling becomes more negative whereas if ρ is smaller than unity the coupling becomes less negative, though never becoming positive, because if ρ is zero $\mathit{f}_a = \frac{a'}{a''}$.

Positive tracer coupling may, however, under special conditions be expected for a divalent carrier system, i.e., if the carrier transported two particles of the same species at a time. Under these conditions we would have a kind of cotransport between the particles of the same species, which may come about in two ways. Either because the binding of one particle to one site increases the affinity of the other site, and vice versa, or because the doubly loaded carrier moves faster than the singly loaded one. This would lead to "positive" tracer coupling, at least in the low concentration range, when competition is very low. If the solute concentration rises toward saturation ranges, the coupling would become negative, as in ordinary competition.

To illustrate this, we simply assume that only the carrier with the two ligands (xa_2) is able to pass the membrane appreciably. The flux ratio should be

$$f_a = \frac{a'^2 \, (K_a^2 + a''^2)}{a''^2 \, (K_a^2 + a'^2)}$$

f_a is obviously greater than (a'/a''), if $a' > a''$. With increasing concentrations of a' and a'', this deviation of f_a from the activity ratio will decline and, as the concentrations tend toward saturation, invert.

The effects of all kinds of isotope interactions on the flux ratio will vanish if $a' = a''$; then also f_a becomes unity so that the flux equation holds. This is in contrast to the corresponding effects of coupling to nonconjugate flows, which can usually not be abolished by making $a' = a''$. It is clear from these examples that before inferring active transport from inequality between flux ratio and activity ratio, isotope interaction has to be either excluded, or corrected for.

4.2.4 Solvent Drag: In the previous model, we have disregarded any solute-solvent interaction during the penetration process. In the absence of the required restraints, however, as is the case in many biological systems, the movement of solutes across the membrane is likely to entail bulk movement of solvent across the membrane.

Any bulk movement must in turn affect the opposing unidirectional solute fluxes in such a way that the flux ratio deviates from the activity ratio. These deviations are similar to those in isotope interaction only if the bulk flow is due to a difference in osmolarity of the solute under consideration. Only then should the difference between flux ratio and activity ratio of this solute vanish if the latter becomes unity. Since such a situation is rarely encountered in biological systems, isotope interaction due to solvent drag has presumably no real significance. To the extent, however, that bulk flow is due to nonconjugate forces, for instance, to differences in hydrostatic pressure or in the osmolarity of different solute species, we should have true coupling, namely, between solute flow and solvent flow, which will be discussed in Chapter 8.

5 Energetics of One-Flow Systems
Treatment of One-Flow Systems in Terms of Thermodynamics of Irreversible Processes (TIP)

5.1 Free Diffusion

We shall now apply the thermodynamic treatment to the same processes as described above using the same simplified models as those serving to illustrate the treatment according to the LMA. The flow of a single solute, in the absence of nonconjugate forces, should be proportional to the conjugate driving *force*, and inversely proportional to a resistance coefficient. The force is conventionally defined as the gradient of the electrochemical potential of the solute i, $-\frac{d\tilde{\mu}_i}{dx}$, and the resistance coefficient, r_{ii}, as the resistance of unit area and per unit distance perpendicular to the membrane area at the point x of the membrane

$$J_i \cdot r_{ii} = -\frac{d\mu_i}{dx} = -RT \frac{d\ln\tilde{a}_i}{dx} \tag{5.1}$$

$\tilde{a}_i$ being the "electrochemical activity" of solute i. Since in the steady state the flow i is constant throughout the membrane, we can integrate Eq. (5.1) over the whole thickness (δ) of the membrane; we do so first under the assumption that i is non-ionic, so that we can disregard electric effects for the moment.

$$- \int_0^\delta d\mu_i = J_i \int_0^\delta r_{ii} dx \tag{5.2}$$

The term $\int_0^\delta r_{ii} dx$ can be replaced by R_{ii}, the "integral specific resistance" of the membrane to solute i, so that we obtain $-\Delta\mu_i = R_{ii} J_i$. Since $-\Delta\mu_i$ is usually replaced by X_i, we obtain

$$X_i = R_{ii} J_i \tag{5.3}$$

Since $J_i = L_{ii} X_i$, L_{ii} being the phenomenological coefficient, it follows that for free diffusion, i.e., in the absence of coupling,

$$L_{ii} = \frac{1}{R_{ii}} \tag{5.4}$$

A precise determination of R_{ii} in terms of concentration and diffusion coefficient of the solute concerned is possible for a homogenous membrane phase, in which D_i would be constant throughout. The gradient of the solute within the membrane would then be linear:

$$\frac{dc_i(m)}{dx} = \text{const.} = \frac{\Delta c_i}{\delta} \tag{5.5}$$

δ being the thickness of the membrane.

At any point of membrane phase the resistance coefficient r_{ii} is inversely proportional to the local concentration times the mobility of the solute concerned (u_i),

$$u_i = \frac{D_i}{RT}$$

$$r_{ii}(x) = \frac{1}{u_i \cdot c_i(x)} = \frac{RT}{D_i \cdot c_i(x)} \tag{5.6}$$

In the steady state, and if u_i is constant throughout the membrane, we obtain by integration of r_{ii} dx over the thickness (δ) of the membrane, i.e., between $x = o$ and $x = \delta$,

$$R_{ii} = \int_0^\delta r_{ii} \, dx = \frac{\delta}{u_i} \frac{\Delta \ln c_i}{\Delta c_i} \tag{5.7}$$

$$R_{ii} = \frac{RT}{D_i} \delta \frac{\Delta \ln c_i}{\Delta c_i} \quad \text{or} \quad = \frac{RT}{P_i} \frac{\Delta \ln c_i}{\Delta c_i}. \tag{5.8}$$

If (c_i'/c_i'') is close enough to unity, let us say between 0.3 and 3, $\frac{\Delta \ln c_i}{\Delta c_i}$ can be replaced by $\bar{c}_i$, the arithmetic mean of the two concentrations with an error not greater than 10%. Even though developed for a homogeneous membrane, the last equation also holds for varying D_i, provided that $\frac{1}{P_i} = \int_0^\delta \frac{dx}{D_i}$.

Obviously, $-\Delta\mu_i$ no longer fits the rigorous physical definition of a force. Its dimension is rather that of an affinity, here of the effort required to transfer one mole of i from compartment 1 to compartment 2. Accordingly, the integrated equation lends itself to the operational treatment of transport in analogy to a chemical reaction (quasi-chemical treatment), which we shall advantageously apply to the more complicated system of coupled flows farther below. At this juncture we may illustrate the close analogy between chemical and osmotic processes by comparing a simple chemical reaction with an equally simple osmotic process. We assume a chemical reaction in which ν_a molecules of A are reversibly transformed into ν_b molecules of B, according to the equation

$$\nu_a A \rightleftharpoons \nu_b B.$$

γ_i, the stoichiometric coefficients, are by convention negative on the left, and positive on the right side of the reaction equation.

The *effort*, here called "affinity" (A_{ch}), which drives the reaction from left to right is

$$A_{ch} = - \Sigma \nu_i \mu_i = - (-\nu_a \mu_a + \nu_b \mu_b) \qquad (5.9)$$

We now apply the analogous treatment to an osmotic process in which γ_a moles of solute A are translocated through a membrane from side ' to side ", according to the equation

$$\nu_a^{'} A' \rightleftharpoons \nu_a^{''} A''$$

Stoichiometric coefficients different from unity are to indicate that in this process the solute A is translocated as a complex consisting of γ molecules of A, rather than as a single molecule. If we treat the translocatable complex on the ' side as the reactant and that on the " side as the product of the process, we can define a driving effort (A_{osm}), analogously to that of the chemical reaction:

$$A_{osm} = -\Sigma \nu_i \mu_i = - (-\nu_a^{'} \mu_a^{'} + \nu_a^{''} \mu_a^{''}) . \qquad (5.10)$$

Since $\gamma_a^{'}$ and $\gamma_a^{''}$, though of different sign, are numerically equal in this case,

$$|\nu_a^{'}| = |\nu_a^{''}| = \nu_a$$

Eq. (5.10) can be written

$$A_{osm} = - \nu_a \Delta \mu_a \qquad (5.11)$$

or, replacing, as is customary, $- \Delta \mu_a$ by X_a

$$A_{osm} = \nu_a X_a \qquad (5.12)$$

The transport equation, under the assumption that solute A penetrates the membrane as a monomer ($\gamma_a = 1$), is accordingly

$$J_a = L_{aa} X_a \qquad (5.13)$$

It is identical with Eq. (5.3).

Eq. (5.13), by describing the net flux of a single species as a linear function of a chemical potential difference, seems to contradict the permeability equation derived in the previous chapter, which describes the same flux as a linear function of a concentration or activity difference. Clearly this flux cannot be a linear function of both the difference between two concentrations and the difference between the logarithm of these concentrations. To the extent that the LMA-parameter P_i can be treated as a constant, or as being at least independent of the concentrations, the corresponding TIP parameters, R_{ii} or L_{ii}, strictly cannot be constant. It may, however, be so treated by an approximation, which is permitted under certain conditions only. This limitation is a serious drawback for the application of the otherwise so convenient notation of irreversible thermodynamics. Before trying to define these conditions, let us investigate how the resistance parameters r_{ii} and R_{ii} relate to the corresponding diffusion parameters D_i and P_i.

According to Fick's law [Eq. (2.1)]

$$J_i = -D_i \frac{dc_i}{dx} \quad , \quad J_i \text{ being } \frac{dn_i}{dt} \frac{1}{A}$$

whereas according to Eq. (5.1)

$$J_i \, r_{ii} = -RT \frac{d \ln c_i}{dx} = -\frac{RT}{c_i} \frac{dc_i}{dx}$$

From these equations we obtain

$$r_{ii} = \frac{RT}{D_i \cdot c_i} \tag{5.14}$$

Since $\frac{D_i}{RT} = u_i(x)$, i.e., the mobility of solute i within the membrane phase in this region, r_{ii} is inversely proportional to the product of mobility $u_i(x)$ and to the concentration $c_i(x)$ of the solute of the membrane phase at distance x from the surface.

In order to relate the corresponding integral parameters (P_i and R_{ii}) to each other we recall from Eq. (2.3) that

$$J_i = -P_i \, \Delta \, c_i$$

In the steady state J_i is constant over the whole membrane thickness (δ) so that at any value of x

$$P_i \, \Delta \, c_i = D_i \frac{dc_i}{dx} \tag{5.15}$$

Evaluating D_i from this equation and inserting it in Eq. (5.14) we obtain

$$r_{ii} \, dx = \frac{RT}{P_i \Delta c_i} \cdot \frac{dc_i}{c_i} \tag{5.16}$$

Integrating this expression over the whole thickness of the membrane (δ) we obtain the integral resistance as in Eq. (5.8)

$$R_{ii} = \int_0^\delta r_{ii} \, dx = \frac{RT}{P_i} \frac{\Delta \ln c_i}{\Delta c_i}$$

R_{ii} in contrast to r_{ii}, is thus a function of the chemical potential difference. In a rather narrow range, however, $\frac{\Delta c_i}{\Delta \ln c_i}$, the logarithmic mean of c_i can be replaced by $\bar{c}_i$, the arythmethic mean of c, so that the phenomenological coefficient can be treated as dependent on $\bar{c}_i$ only. This is possible to the extent that $\Delta \ln c_i$ can be replaced by the approximation based on the series

$$\ln x = 2 \left[\frac{x-1}{x+1} + \frac{1}{3} \frac{(x-1)^3}{(x+1)^3} + \cdots \right]$$

All terms except the first one can be neglected without causing an error of more than 10 %, if $0.33 < x < 3.0$. Applying this approximation to our system, setting $x = (c_i'/c_i'')$ we obtain

$$\Delta \ln c_i = 2 \, \frac{c_i'' - c_i'}{c_i' + c_i''} = \frac{\Delta c_i}{\bar{c}_i} \quad \text{or} \quad \Delta c = \bar{c}_i \, \Delta \ln c_i \qquad (5.17)$$

Hence:

$$J_i = \frac{P_i}{RT} \, \bar{c}_i \, X_i \qquad (5.18)$$

$$R_{ii} = \frac{RT}{P_i \, \bar{c}_i} \, , \quad \text{and} \quad L_{ii} = \frac{P_i \, \bar{c}_i}{RT}$$

To the extent that $\bar{c}_i$ is constant we may therefore expect a fairly linear relationship between J_i and X_i within a range in which the above approximation is permissible.

Experimentally it is often not practicable to keep $\bar{c}_i$ constant, especially not in studies of transport through cellular membranes Usually one tries to maintain c_i in one phase on a fairly constant level while varying c_i in the other phase. Consequently $\bar{c}_i$ will vary and further narrow down the linearity of the above relation. This is illustrated by Figure 10 in which the rate J_i, as predicted from the kinetic equation, is plotted versus the corresponding X_i values. X_i is varied by varying c_i''. It can be seen that only within the range $X_i = \pm 0.2 \, RT$ is the deviation from linearity 10 % or less. Since most studies on biological transpor deal with driving efforts much higher than $\pm 0.2 \, RT$, the method of TIP is clearly unsuitable for describing free diffusion. Fortunately the range of linearity can be considerably extended with saturable systems, as we shall see later, by selecting suitable experimental conditions. In such systems the dependence of R_{ii} on $\Delta \mu$ appears to be counteracted by a dependence on other variations in the system owing to fortunate circumstances. So the application of TIP may be less restricted and more useful than is suggested by some more pessimistic predictions, provided that the range in which phenomenological coefficients, L or R, remain reasonably constant, can be ascertained experimentally.

5.2   Facilitated Diffusion

The question arises whether and in which way the notation of irreversible thermodynamics as in Eq. (5.13), can be meaningfully applied also to facilitated diffusion. Since the conjugate drivinc force in facilitated diffusion is the same as in free diffusion, the two processes should not be different from the energetic poin of view. Kinetically, however, they are different, since in facilitated diffusion the flow is no longer a linear function of the concentration difference, owing to the fact that the number of carrier sites is limited. If, for instance, the carriers are saturated on both sides of the membrane, there should be no appreciable net transport at all, no matter how great the difference of the concentrations of solute between the two compartments. One might expect also that with varying solute concentrations on one or both sides of the membrane, the phenomenological rate

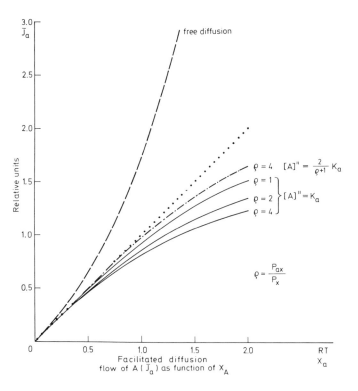

<u>Fig. 10.</u> Linearity of TIP functions in facilitated diffusion. The net flow of an uncharged solute species (*A*) is plotted against X_A, the negative electrochemical potential difference. The rate is derived from the corresponding kinetic equation, under the assumption that the steps involved are first order reactions. The condition is that the concentration of *A* on the trans side is kept constant at the value indicated, whereas the concentration of *A* on the cis side is varied. The *broken line* would obtain if *A* passes the membrane exclusively by free diffusion. The *solid line* would obtain if the solute passes the membrane exclusively by a carrier-mediated transport. ρ gives the ratio of the mobility of the loaded carrier to that of the empty carrier. The function fits the theoretical straight line (*dotted line*) most closely if A'' is equal the half saturation (Michaelis constant

coefficient should also vary greatly over a very large range. In reality, however, saturable systems, etc., may under certain conditions yield a better linear relationship in the plotting in terms of irreversible thermodynamics than does, for instance, free diffusion. This may be illustrated for the simplest model of facilitated diffusion of the solute A, a model which is symmetric and in which the mobilities for the loaded and unloaded carriers are equal. As before, we write the rate equation in terms of the LMA, i.e., as the difference between two opposing saturable rates:

$$J_a = J_a^{max} \left(\frac{a'}{K_a + a'} - \frac{a''}{K_a + a''} \right) = J_a^{max} \frac{K_a(a' - a'')}{(K_a + a')(K_a + a'')} \quad (5.19)$$

In terms of TIP the same rate should be

$$J_a = L_{aa} RT \ln \frac{a'}{a''} \quad (5.20a)$$

or

$$J_a = L_{aa} RT \left(\ln \frac{a'}{K_a} - \ln \frac{a''}{K_a} \right) \quad (5.20b)$$

If the concentrations of a' and a" are chosen to be close enough to K_a, e.g., $< 3 K_a$ and $> \frac{1}{3} K_a$ the known approximation to convert the logarithmic expression can be used [Eq. (5.17)], so that Eq. (5.20b) changes to

$$J_a = L_{aa} RT \cdot 2 \left(\frac{a' - K_a}{a' + K_a} - \frac{a'' - K_a}{a'' + K_a} \right) \quad (5.21)$$

$$J_a = L_{aa} RT \frac{4 K_a (a' - a'')}{(K_a + a')(K_a + a'')}$$

which is identical to Eq. (5.19) above, if we set

$$4 L_{aa} RT = J_{max}$$

$$L_{aa} = \frac{J_{max}}{4 RT}$$

Accordingly the phenomenological coefficient is here seemingly a true constant, but only within the range in which the above approximation holds, namely, between $3 K_a$ and $\frac{1}{3} K_a$. This range covers about $2 RT$, i.e., 1.2 kcal/mol and is hence five times wider than the corresponding one for free diffusion.

In the following, we shall convert the logarithmic expression precisely, i.e., without the above approximation; we start again with Eq. (3.8):

$$J_a = \frac{P_x X_T}{2} \frac{K_a(a' - a'')}{(K_a + a')(K_a + a')}$$

To transform this equation into a TIP equation, we have to multiply numerator and denominator by $RT \, \Delta \ln a$. Hence

$$L_{aa} = \frac{P_x X_T K_a \cdot (a' - a'')}{(K_a + a')(K_a + a'') \, \Delta RT \ln a} \quad (5.22)$$

which for the case that a" = K_a becomes

$$L_{aa} = \frac{P_x X_T (a' - K_a)}{2(K_a + a') RT \, \Delta \ln a} \quad . \quad (5.23)$$

This is the real phenomenological coefficient, which is no longer limited by any approximation. It is clearly not independent of a'. This dependence is, however, very small, as can be seen from the accompanying graph (Fig. 10). If we complicate the model by introducing different rate coeffcients for the loaded and unloaded carrier, respectively, $(P_{ax} \neq P_x, \rho = \frac{P_{ax}}{P_x} > 1)$, the dependence of L_{aa} an a', and hence the deviation of the rate function from linearity, becomes greater. The graph shows the transport rate J_a, as determined from Eq. (5.19) but plotted as a function of the chemical potential difference. It can be seen that with $\rho = 4$ the 10 % limit of the deviation has already been passed if $\Delta\mu$ exceeds RT. It can also be seen, however, that the good fit of the simple equation can be recovered if instead of K_a the Michaelis constant K_m is used, i.e., if

$$a'' = K_m = K_a \frac{2}{1 + \rho}.$$

By introducing asymmetry, the relationship between the two representations, by LMA and TIP, respectively, becomes more involved by the introduction of more different parameters. The linearity of the phenomenological equation, however, is here also better than with free diffusion.

5.3 Tracer Coupling

To treat tracer coupling in terms of TIP one cannot use unidirectional fluxes, by which isotope interaction is indicated, since they are thermodynamically meaningless. This is because a unidirectional flux is measured by adding the tracer at only one side of the membrane. Hence the concentration of the label on the trans side must be zero or at least negligible, so that the driving force of the tracer flux, namely, the electrochemical potential difference of the labeled isotope between the two sides of the membrane, is theoretically infinite, or at least thermodynamically indefinable. The only reasonable way to implement unidirectional tracer flows into a thermodynamic treatment is by way of the *flux ratio*, i.e., by the ratio of the forward and backward unidirectional fluxes, measured separately under identical conditions of the system as to overall concentration, temperature, etc., or, more accurately, by using labeled isotopes on both sides.

$$\boldsymbol{f}_i = \frac{\vec{J}_i}{\overleftarrow{J}_i}$$

It has been stated before that for free diffusion, i.e., in the absence of coupling and isotope interaction, Ussing's flux equation is valid. It states that the flux ratio equals the activity ratio, i.e., the ratio of the electrochemical activities of the solute at the two sides of the membrane.

$$\boldsymbol{f}_i = \frac{\tilde{a}_i'}{\tilde{a}_i''} \tag{5.24}$$

It has been derived by KEDEM and ESSIG (1965) that in the absence of isotope interaction

$$RT \ln \boldsymbol{f}_i = J_i \, R_{ii} \tag{5.25}$$

R_{ii} being the integral net resistance and J_i the net flow of solute i.

This equation holds also in the presence of true coupling, e.g., to the flow of solute i, in which case

$$R_{ii} \, J_i = X_i - R_{ij} \, J_j$$

R_{ij} being the integral cross coefficient.

Clearly only in the absence of such coupling is the flux ratio equal to the electrochemical activity ratio, because then

$$R_{ii} \, J_i = X_i = RT \ln \frac{\tilde{a}_i'}{\tilde{a}_i''}$$

in agreement with Ussing's flux equation $\left[\text{Eq. (2.7a)} \right]$. To the extent that this equation holds, it yields the integral net resistance, R_{ii}, from the flux ratio and the net flux of i, or, since

$$J_i = \overrightarrow{J}_i - \overleftarrow{J}_i$$

from the net flux J_i and a single unidirectional flux and without knowledge of the activity ratio.

However, as has been stated before, Ussing's flux equation no longer holds in the presence of isotope interaction not even for uncoupled flows: The flux ratio may then greatly differ from the activity ratio. Accordingly the R_{ii} value obtained by the above equation would differ from the true integral resistance. To distinguish this resistance term based on the flux ratio from that based directly on the driving force, one calls it "exchange resistance" (R_{ii}^*), as defined by the following equation

$$RT \ln \boldsymbol{f}_i = R_{ii}^* \cdot J_i \tag{5.26}$$

or

$$RT \ln \boldsymbol{f}_i = \frac{1}{L_{ii}^*} \cdot J_i$$

According to KEDEM and ESSIG (1965)

$$R_{ii}^* = \int_0^\delta (r_{ii} - r_{ij}) \, dx \quad \text{and} \quad R_{ii} = \int_0^\delta r_{ii} \, dx \tag{5.27}$$

r_{ij} being the cross resistance coefficient relating the electrochemical potential gradient of the translocated solute to the flux of its isotope.

Obviously R_{ii}^* is equal to the true integral resistance (R_{ii}) only if $r_{ij} = o$, i.e., in the absence of isotope interaction. R_{ii}^* is smaller than R_{ii} if r_{ij} is positive (negative coupling) — and greater than R_{ii} if r_{ij} is negative (positive coupling). — It follows that in the absence of true coupling

$$RT \ \ln \boldsymbol{f}_i = R_{ii}^* \ J_i = \frac{R_{ii}^*}{R_{ii}} \ X_i \qquad (5.28)$$

We see that $RT \ \ln \boldsymbol{f}_i$ is related to the true driving force as is R_{ii}^* to R_{ii}, the integral resistances of the isotope flux and of the bulk flow, respectively. As long as the ratio (R_{ii}^*/R_{ii}) is not known, we cannot exclude isotope interaction, and hence cannot postulate the presence of a nonconjugate driving force, i.e., active transport, unless it can be shown that at zero conjugate driving force ($X_i = 0$) the flux ratio is unity. Hence only at $X_i = 0$ can a deviation of the flux ratio from the activity ratio be taken to indicate true coupling. Even then, however, the magnitude of the nonconjugate force cannot be determined from the flux ratio in the presence of isotope interaction, as follows from the equation, expressing the coupling between the flows of solute i with that of solute k:

$$RT \ \ln \boldsymbol{f}_a = \frac{R_{ii}^*}{R_{ii}} \ (X_i - R_{ik} \ J_{ik}) \qquad 5.29)$$

which should hold if the flow of solute i is truly coupled to the flow J_k.

It may be illustrative to relate the treatment of isotope interaction in terms of TIP to the treatment previously carried out in terms of the LMA. This may be done for our previous models by deriving, for instance, the ratio (R_{aa}^*/R_{aa}) from the kinetic equations of the penetration of solute A:

$$X_a = R_{aa} \cdot J_a$$

and

$$RT \ \ln \boldsymbol{f}_a = R_{aa}^* \ J_a \ , \ \left[\text{Eq. (5.26)}\right]$$

$$\frac{R_{aa}^*}{R_{aa}} \ \text{should be} \ \ \frac{RT \ \ln \boldsymbol{f}_a}{X_a} = \frac{\ln \boldsymbol{f}_a}{\ln \dfrac{a'}{a''}} \qquad (5.30)$$

To apply this relationship to the kinetic equations we have to get rid of the logarithmic expressions, using the well-known approximation $\left[\text{Eq. (5.17)}\right]$

$$\ln x = 2 \ \frac{x-1}{x+1} \ + \ \ldots$$

So we obtain for those cases of tracer coupling that we treated previously:

1. Single-file (or channel-mediated) diffusion, assuming P_t to be very great

$$\frac{R_{aa}^*}{R_{aa}} = \frac{(a' + a'' + 2K_a) \ (a' + a'')}{(a' + a'' + 2K_a) \ (a' + a'') - 2a'a''} \qquad (5.31)$$

2. Simple dimer formation

$$\frac{R_{aa}{}^*}{R_{aa}} = \frac{2P_{aa}(a' + a'') + P_a K_a}{2P_{aa}(a' + a'') + P_a K_a - 4P_{aa}\frac{a' \cdot a''}{a' + a''}} \tag{5.32}$$

In both cases $\frac{R_a{}^*}{R_a} > 1$ as is characteristic of *positive* tracer coupling.

3. For facilitated diffusion

$$\frac{R_{aa}{}^*}{R_{aa}} = \frac{K_a(a' + a'')}{K_a(a' + a'') + 2 \rho\, a'\, a''} \tag{5.33}$$

$\frac{R_{aa}{}^*}{R_{aa}}$ is < 1 as would be expected for *negative* tracer coupling.

It is interesting to note that if in isotope interaction the activity ratio, and as a consequence the flux ratio become unity, $(R_{aa}{}^*/R_{aa})$ does not become unity.

We see that in the notation of irreversible thermodynamics the phenomena resulting from isotope interaction, such as exchange phenomena, appear only implicitly in the exchange resistance $R_{aa}{}^*$, which, however, can only be larger or smaller than the true resistance but cannot give any information about the special type of interaction.

Two-Flow Systems –
Energetic Coupling

6 Solute-Specific Coupling – Active Transport

6.1 Mechanistic Aspects of Coupling

6.1.1 Criteria and Definitions. In the previous chapters, the mechanism of transport has been discussed without considering any input of energy from sources other than the *conjugate driving force*, i.e., other than the electrochemical potential gradient of the transported solute itself. In other words, only *passive* transport processes have been considered, which come to an end as soon as the moving species has reached equilibrium distribution, i.e., the same electrochemical potential on both sides of the membrane. In many biological transport systems, however, it is observed that the net movement of some solutes goes for beyond this equilibrium distribution, coming to an end only if a considerable electrochemical potential difference has been erected. In other words, this extra transport is energetically "uphill", and is therefore called "active". Active transport can occur only if the movement of the transported species is driven by an outside force. In other words, an additional *nonconjugate* driving force must be effective in driving the flow of the species concerned against its electrochemical potential gradient. For this to occur the "driven" flow has to be energetically *coupled* to some other, energy-releasing process, conveniently called "driver" process. Active biological transport, since it draws its energy largely from metabolism, must be ultimately coupled to respiration or fermentation. This does not mean that the transport is coupled directly to either the oxygen-consuming or to a glycolytic reaction, respectively; rather many intermediate steps with the transfer of energy appear to be involved. The crucial question in this context is to what reaction or process the translocation is coupled *directly*. In view of the universal role of energy-rich phosphates in energy metabolism, it was previously assumed that all active transport processes are directly coupled to the hydrolysis of energy-rich phosphates. Indeed, there is evidence that energy-rich phosphates, ATP, PEP, etc., are required for, and utilized in, active transport. More recently, however, another process is considered to function as an immediate energy source for many transport systems, namely, the movement of a cation down its electrochemical potential gradient. A lot of evidence favors this view, H ions being the main ions for microbial systems, and Na ions for the cytoplasmic membrane of animal cells. Before trying to answer these questions we shall discuss some general problems concerning the definition and experimental verification of active transport.

The effect of a nonconjugate force on the transport of a given solute may manifest itself in various ways: It may, for instance, cause the transportee to "accumulate", i.e., to flow against its

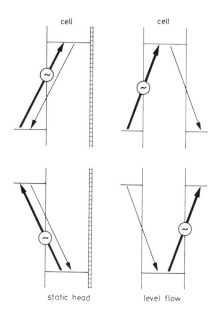

Fig. 11. Static head and level flow. Pump-leak systems between cell and surroundings. In static head either the inward movement or the outward movement may be active whereas the other is passive. In level flow the net movement of solute is from left to right across the cell. Also here either the inward movement or the outward movement may be active

static head   level flow

(electrochemical) activity gradient. In the absence of an opposing activity gradient the nonconjugate driving force should *accelerate* the net flow of the transportee in the direction of transport. Either effect is often difficult to verify. The accumulation effect implies that the activity of the transportee is higher on the trans side than on the cis side of the transporting membrane. Evidence to this point is often difficult to supply, especially if the activity of the transported ion in one of the compartments cannot be precisely determined. On the other hand, an accelerative effect in one direction is also difficult to verify since it might be confused with retification phenomena that may occur in asymmetric systems of facilitated diffusion, as has been discussed in Chapter 3. Sources of error are largely evaded, if the coupling is studied under two special conditions: *level flow* and *static head* (Fig. 11), which will now be characterized.

In *level flow* the electrochemical activities of the transported solute are equal on both sides of the membrane. In order that the net movement does not disturb this condition, the compartments must be large, theoretically infinite. Since the conjugate driving force is then zero, level flow is driven exclusively by nonconjugate forces. In biological systems level flow is approximately verified when large amounts of solutes are reabsorbed or excreted without an appreciable concentration gradient, such as in transepithelial transport in the intestine and in the kidney tubules. In short: level flow is the net transport at zero difference in activity of the transported solute $(J_a)_{\Delta a=0}$. *Static head*, on the other hand, requires the existence of   at least one finite compartment, in which the activity of the transportee will rise, or fall, respectively, until a state is reached in which net transport stops because the rate of leakage has become equal to the intrinsic rate of active transport. Accordingly, static head can be characterized as the activity ratio at zero net flow $\left(\frac{a''}{a'}\right)_{J_a=0}$.

Each condition, level flow and static head, can be used to test for active transport. At vanishing concentrations of transported solute, the flux ratio at level flow and the static head value should give the same value, which is identical with the *Haldane ratio*. The Haldan ratio, as discussed elsewhere, is the ratio of the maximum initial net flow of the transported solute to the Michaelis constant of this flow, divided by the corresponding ratio of the initial net flow in the opposite direction (p. 100):

$$HR = \frac{\overrightarrow{J}_a^{max}}{K_m'} \quad \frac{K_m''}{\overleftarrow{J}_a^{max}} ; \tag{6.1}$$

In the absence of a nonconjugate driving force, the Haldane ratio should be unity, for a symmetric system as well as for an asymmetric system. Hence the Haldane ratio, determined by any of the usual methods, should be a useful means to test for, and assess the effectiveness of, active transport, experimentally in a real system, or theoretically in a model system.

The *Haldane ratio* can be satisfactorily obtained for transport between two bulk phases of known composition, pressure, and electrical PD, in which the electrochemical activities of the transported solutes are equal, or at least known. However, difficulties arise if one of the compartments is a cell or cellular organelle, in which the activities of solutes are obscured by various factors. If, for instance, a solute is adsorbed or bound to some intracellular material, its activity is bound to be overestimated. If, on the other hand, the cell contains some compartments that do not admit the transported solute, the activity of this solute in the accumulating compartment is bound to be underestimated. In the first case active transport may be postulated where there is none, whereas in the later case existing active transport may be overlooked. In either case it is helpful to test for another criterion of active transport, namely the dependence on metabolic activity of the cell, for instance, by the use of metabolic inhibitors. The observation that such inhibition does not affect the transport of a given solute would argue against active transport, since dependence in metabolism is a necessary, though not sufficient, criterion of active transport. It is not sufficient since by itself it does not prove active transport, especially not of an ion that is driven by an electrical PD. The maintenance of an electrical PD requires an active process dependent on metabolism. Consequently, the depression of the electrical PD by metabolic inhibition may also affect the movement of purely passive ions.

In summary, the demonstration that the movement of a solute through a membrane is stoichiometrically coupled to another process and thus subject to a nonconjugate driving force, should suffice to prove active transport, but can often not be verified for the above-mentioned reasons. In some such cases the nonconjugate driving force can be assessed with the help of metabolic inhibition, to the extent that it specifically abolishes this force. This procedure is likely to help in transport systems for uncharged solutes, but is ambiguous in ion transport since, as discussed above, metabolic inhibition may affect the electrical PD and thus also the conjugate driving force as well.

6.1.2 Driving Forces. In order that forces other than the conjugate force become effective in driving a given species (i) across a membrane or barrier, such as in active transport, the flow of i must be coupled to other flows. Some of the free energy liberated by the latter may then become available to the former, thereby driving it against its conjugate driving force. Coupling may, on the one hand, occur between the *flows* of two species moving simultaneously across a membrane or barrier, in the same direction (cotransport or symport) or, under special conditions, in opposite directions (countertransport or antiport). These kinds of coupling we may call *"osmo-osmotic"*, since it is between two osmotic processes. Coupling may, on the other hand, also occur between the flow of a species across a membrane and the flow of a chemical reaction, taking place within the membrane. This coupling, which we shall call *"chemi-osmotic"* [7], in its most rigorous definition was formerly considered the only basis of true active transport. Obviously, however, our previously stated criteria of active transport can, under certain conditions, also be verified in osmo-osmotic coupling. If the flow of the driving species depends on metabolism, e.g., through active transport, then also the flow of the driven species, to which it is coupled, will depend, though indirectly, on metabolism. If, for instance, the uphill flow of one solute (A) is coupled to the downhill flow of another solute (B), and if the gradient of B is maintained by (primary) active transport of B, then not only the flow of B, but indirectly also that of A will depend on metabolic energy supply. It is therefore justified to call the transport of A active, but since its coupling to a chemical reaction is only indirect, via the flow of B, we call it "secondary active", in contrast to transport via chemi-osmotic coupling, which we call "primary active". Thus, expanding the formerly too-restrictive definition of active transport to include secondary active transport appears to be recommendable also from the practical point of view, because in many biological transport systems it has been very difficult, or even impossible, experimentally to decide whether the transport is primary or secondary active.

Before dealing with the various types of coupling that have so far been postulated for biological transport systems, we shall first discuss some underlying mechanistic principles. Coupling between two processes with the transfer of energy requires that the two processes are linked so tightly that the one cannot, or at least cannot optimally, go on without the other, and vice versa. Within cellular metabolism many coupling processes have

---

[7]The term chemiosmotic is used here in a wider sense that is commonly used, which restricts this term to a special case of coupling, namely, that between ion flow and ATP synthesis, according to the hypothesis of Mitchell.

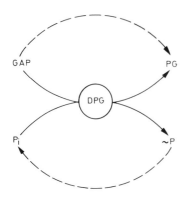

Fig. 12. Coupling between glyceraldehyde phosphate (*GAP*) dehydrogenation and ATP synthesis via diphosphoglycerate (*DPG*). For details see text

been observed and described. They are of great importance in enabling *endergonic* reactions, i.e., those in which free energy is taken up, to take place at the expense of *exergonic* reactions, i. e., those by which free energy is liberated. The underlying principle of biochemical coupling is usually characterized as the principle "of the common intermediate". This principle can be derived from one of the first elucidated cases of coupling, that between the (exergonic) dehydrogenation of glyceraldehydephosphat (GAP) to 3-phosphorglycerate (3-PG) and the (endergonic) synthesi of ATP from ADP and inorganic phosphate (P_i), the 1,3-diphosphoglycerate (DPG) being the crucial intermediate by which the two above reactions are forced to proceed jointly through the following sequence of reactions:

$$GAP + P_i + NAD^+ \longleftrightarrow DPG + NADH$$
$$\underline{DPG + ADP \qquad \longleftrightarrow 3\text{-}PG + ATP}$$

overall: $GAP + P_i + ADP + NAD^+ \longleftrightarrow 3\text{-}PG + ATP + NADH$

We see that this common intermediate (DPG) ties two seemingly unrelated reactions together: The oxidation of P-glyceraldehyde to P-glycerate with the transformation of an inorganic (P_i) to a phosphoryl group ($\sim P$) of the energy-rich ATP (Fig. 12).

This coupling is "energetic", i.e., associated with the transfer of energy between the two coupled reactions: The energy liberated by the exergonic oxidation is used to drive the endergonic synthesis of ATP.

If the formation of the common intermediate is the only pathway by which each reaction is allowed to proceed, the coupling is absolutely tight, i.e., it is 100 %. This tightness will, however, never be attained in reality. There will always be some possibility for each reaction to go by a route by-passing the common intermediate, as indicated by the dotted lines: For instance, the oxidation of GAP may to a small extent occur directly, and also some ATP may hydrolyze without the intermediate formation of DPG. These by-passing routes are called "leakages", which make the coupling less tight and reduce the efficiency of energy transfer.

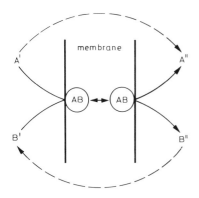

Fig. 13. Coupling between the penetration of two different solutes, A and B, across the membrane via a penetrating complex between the two (AB). For details see text

The principle supposedly underlying the coupling in primary and secondary active transport is analogous to that of the "common intermediate". For reasons that will become clear later, we prefer in this context to call it the principle of the "common step". It can be most simply exemplified by the coupling between the flows of two solutes, A and B, across a membrane. Such coupling will occur if the membrane is poorly permeable to each A and B alone but well permeable to a combination of them, AB. In this case the formation and penetration of AB would be the common step that links the seemingly unrelated steps to each other: The transport of solute A and that of solute B (Fig. 13). Physiological examples of this kind of coupling may result from the formation of neutral or lipophilic molecules from electrolyte ions, such as in acid base reactions or in complex formation. So the CO_2, formed from HCO_3 in the reaction:

$$HCO_3^- + H^+ \rightleftharpoons H_2O + CO_2$$

owing to its lipophilic nature, penetrates the membrane much faster than do either HCO_3^- or H^+ alone. Accordingly, the penetration of bicarbonate, for example, through the kidney tubules membrane, is found to be greatly stimulated by acidification of the medium. Similarly, the reabsorption of various drugs by the intestine is also promoted by acidification, which converts the anions into electrically neutral acids. Analogous is the formation of more permeant neutral amines from cationic ammonium ions by OH-ions. Also here the coupling may be energetic, i.e., associated with energy transfer. If the movement of A is exergonic, i.e., energetically downhill, e.g., if a' > a", the energy liberated by this movement may, at least partly, be used to drive the (endergonic) energetically uphill transport of B. Also here the coupling is likely to be less than 100 %, owing to leakages, here caused by some penetration of the membrane by A and B in the uncombined form, as indicated by the dotted lines. The quantitative treatment of leakage and its effect on the tightness (or degree) of coupling and the efficiency of energy transfer will be dealt with in Chapter 7.

The simple kind of coupling just outlined is observed in biological membranes, which are often more permeable to an undissociated compound than to ions formed from it. Even though under such con-

ditions a downhill gradient of a cation may drive an anion energetically uphill, this kind of direct coupling is not the one
usually postulated to occur in active transport. For true active
transport it is tacitly assumed that the linkage between transport and the driver process, be it osmotic or chemical, is indirect, i.e., mediated by a transport carrier, and that it is
the movement of this carrier that is coupled to the driver process. Accordingly active transport can in many cases be considered a coupled form of mediated transport, such as facilitated
diffusion, since it appears to have all those features that are
considered characteristic of carrier-mediated transport, such
as saturation kinetics, etc. There are cases reported in which
by blocking the coupling an active transport system can be transformed into one of facilitated diffusion (WILSON et al., 1970).
We may therefore conveniently base the discussion of the various
models of active transport on the simplified model of facilitated
diffusion presented in the preceding section.

6.1.3 General Model of Active Transport: In this model, the mediated
transport process can be regarded as a cyclic process consisting
of at least four different steps in series:

1. Binding
2. Translocation
3. Release
4. Relocation

Translocation and relocation (steps 2 and 4) appear to be vectorial processes, i.e., directed in space, whereas binding and
release reactions are usually treated as scalar processes taking
place in or at the interfaces between the membrane phase and the
adjacent solutions.

To transform this system into one of active transport, we have t
postulate that it is coupled to another process, for convenience
called here the "driver process", which may be the transport of
another solute (osmo-osmotic coupling), or a chemical reaction
(chemi-osmotic coupling). In contrast to the direct coupling
discussed above, the "common step" need not be the translocation
of A (step 2). In a cyclic process like this, consisting of several distinct steps in series, any of these steps could be the
coupled one, i.e., the one shared with the driver process (JAC-
QUEZ, 1961). A cycle in the steady state may be looked upon as
a wheel: To make it turn, it should be unimportant at which part
of the circumference the force inserts.

6.1.4 General Principles of Coupling. To understand the fundamental
principles of coupling in active transport, we may try to clas-
sify the principles into different and distinct types. However,
this is very difficult, partly because we have to do this in
terms of transport models that are primarily hypothetical rather
than based on solid evidence. This difficulty, however, is not
altogether discouraging to the extent that the principles are
fundamental and hence probably independent of the detailed molecular mechanism of the transport system. There is, however,
a still greater difficulty, namely, to select unambiguous criteria on which the classification may be based. So it happens
that two systems of coupling belong to the same group according

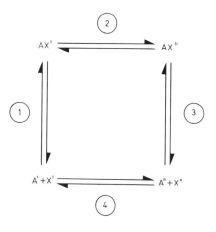

Fig. 14. General model for transport between two solutions. The fundamental four different steps

to one criterion, and to different groups according to another. Hence any classification is necessarily somewhat arbitrary.

One possible basis of classification would be whether the coupling is *direct* or *indirect*, i.e., whether the translocation step of the transported solute is coupled directly to the driver reaction or not. Clearly only coupling at step 2 (Fig. 14), the translocation of the solute, could be called direct, whereas coupling at any other step of the cycle would be indirect. This kind of distinction may be a useful basis for an experimental approach, because with an indirectly coupled system, the translocation should be separable, conceptionally or even experimentally, from the coupling reaction, whereas this is not likely to be so in direct coupling where any translocation may come to a halt whenever the driver reaction stops.

On the other hand, this basis does not appear to be very discriminative since many heterogeneous mechanisms of coupling are indirect. We therefore would like to refine this classification by including the distinction as to whether the coupling is scalar or vectorial. Clearly the reactions concerned with activation and inactivation of the carrier, and hence with binding and release of the solute, can be, and usually are, treated as scalar reactions. Hence we shall speak of scalar coupling whenever either step 1 or step 3 is the common step, i.e., the step shared with the driver reaction.

Scalar coupling produces translocation by building up and maintaining an osmotic gradient for the active carrier, which can then move down passively, without direct linkage to the driver reaction. The principle underlying this kind of (active) translocation we may call the *source- and sink principle*. By contrast, translocation of the loaded or unloaded carrier site, whatever its mechanism, is a vectorial process. Accordingly we shall speak of vectorial coupling whenever step 2 is directly involved. In this case the translocation of the carrier solute complex is not driven by an electrochemical potential gradient but by the affinity of the overall chemical reaction, in which driver and driven reaction are tightly linked to each other. The principle of vec-

72

torial coupling which in contrast to scalar coupling, may force
the carrier complex against its own osmotic gradient we may call
conveyor principle. This particular circumstance might be considered
a fundamental criterium to differentiate direct from indirect
coupling: With scalar coupling we would in the steady state expect that the ratio of loaded carrier over the unloaded one is
higher on the cis side than on the trans side (if the transport
is to go from cis toward trans), whereas the opposite should be
expected in direct vectorial coupling.

$$\text{scalar} \quad \frac{ax'}{y'} > \frac{ax''}{y''} \quad \text{(indirect)} \tag{6.2}$$

$$\text{vectorial} \quad \frac{ax'}{y'} < \frac{ax''}{y''} \quad \text{(direct)}$$

This distinction between scalar and vectorial coupling would become ambiguous if the driver reaction were coupled directly with
step 4, the relocation of the (empty) carrier site. Such a coupling would be vectorial with respect to the translocation of
the binding site of the carrier, but not with respect to the
transport of the solute. The latter would be coupled only indirectly and immediately driven by a source-and-sink mechanism,
similar to that in scalar coupling. No unequivocal evidence, however, is available that coupling at step 4 is a real possibility.

Let us now consider scalar and vectorial coupling separately and
in more detail.

6.1.5 Scalar Coupling — The Source-and-Sink-Mechanism. Primary active
transport, to the extent that it operates on the source-and-sink
principle, has many features that are similar to those of secondary active transport. In either case the translocation of the
transported solute is immediately driven by a quasi-osmotic force
the gradient of the carrier solute complex. In either case step
and/or step 3 are involved in maintaining this gradient. In facilitated diffusion, they merely stand for binding and release,
respectively, of transportee at the appropriate carrier sites.
In active transport they may in addition involve processes that
tend to keep the transporting carrier solute activity on the cis
side higher than on the trans side, thus providing the immediate
driving force for the translocation. We can imagine such a mechanism by assuming that the carrier at each side exists in two
states: X, the transport-active, and Y, the transport-inactive
state. Transport-active means capable of transporting the solute
either by binding it more tightly (affinity effect) or by translocating it more rapidly (velocity effect), than do the corresponding inactive forms.

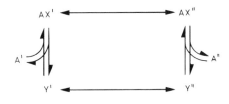

Fig. 15. General diagram of activation-inactivation principle in transport carrier systems. X activated
form, Y inactivated form of the carrier. Activation may occur by binding of cosolute or by chemical alter-
action of the carrier. For details
see text

In cotransport, activation merely means combination with cosolute (B) on the cis side, and in countertransport, combination with countersolute (C) on the trans side

$$Y' + B' \rightleftharpoons X' \quad (= YB) \tag{6.4}$$

$$X'' + C'' \rightleftharpoons Y'' \quad (= XC) \tag{6.5}$$

where B stands for the cosolute and C for the countersolute.

In primary active transport, on the other hand, the transition between X and Y is assumed to involve a chemical change of the carrier produced by an enzymatic reaction that is coupled to an exergonic (driver) reaction. Obviously, for the source-and-sink mechanism to operate, it is of minor importance how the activation, and inactivation, of the carrier comes about, whether the combination with another (co- or counter-)solute, as in co- and countertransport, respectively, or by a chemical or conformational change of the carrier. Essential is that the activation reaction is restricted locally to one side, and the inactivation reaction to the "trans" side; in other words, that reactions 1 and 3 are different from each other, which is achieved by the fact that one is energetically coupled and the other not.

Clearly we are classifying the energetic coupling in secondary active transport as scalar. The reader may wonder whether this is justified, and there are indeed arguments against it. Undoubtedly the penetration of the barrier by the ternary complex represents a common step between the two solute movements. Since each is a vector, we could from this point of view speak of vectorial coupling. On the other hand, secondary active transport, for instance cotransport, could also be envisaged as coming about by the activation and inactivation of the carrier, according to the source-and-sink principle, by binding and release of the cosolute. Both these latter reactions, which are obviously crucial for the whole principle, can be treated as scalar reactions, and hence the coupling could be called primarily scalar. In the present context we would prefer this latter classification, because the immediate driving force for the translocation is the downward electrochemical potential gradient of the carrier-solute complex, rather than that of the cosolute. One may therefore look at secondary transport as if the joint translocation of the two solutes were a secondary effect, incidentally resulting from the scalar reactions at step 1 and step 3. Finally scalar coupling is indicated here by the criterion we have decided to use: The ratio of loaded over unloaded carrier is in the steady state higher on the left side:

$$\frac{xab'}{x'} > \frac{xab''}{x''} \tag{6.6}$$

One could even think of a model of active cotransport of the solute that could be brought about in which driving solute and driven solute pass the membrane for the major part separately. In this model the ternary complex is required to enter the membrane phase but within the membrane there is another barrier that transmits the carrier solute complex (XA) only separately from B, the driver solute. After they have passed this barrier, they may join again in order to leave the membrane phase on the trans side as a ternary complex. This model seems far-fetched, and there is no evidence or even likelihood for its existence, but it shows that for active cotransport the joint movement of solute and cosolute through the barrier is not essential.

Similar, though less acute ambiguities may also appear with scalar coupling. Here, as characteristic for the source- and-sink system, we have put the emphasis on the (scalar) transformation between X and Y, the active and inactive form of the carrier, respectively. The crucial coupling reaction has been described entirely like a scalar reaction, as we know it from metabolic reactions, e.g., for the "push" type on the cis side:

$$Y + S \rightleftharpoons X\text{-}P \tag{6.7}$$

a reaction apparently not containing any vectorial element. We could, however, just as well look at the overall process as at a two-step reaction, the main reaction

1. $Y + S \rightleftharpoons X\text{-}P$ and the after-reaction

2. $X\text{-}P \rightleftharpoons Y + P$

The overall reaction would clearly be $S \rightleftharpoons P$, which because of the fact that the two steps are strictly separated spatially, the first one occurring on the one side only, and the second one, on the other, gets a vectorial feature, i.e., the after-reaction cannot proceed unless preceded by a translocation. This could be illustrated by a hypothetical reaction of the following kind: Let us assume that the active form of the carrier (X) differs from its inactive form by the presence of a hydrolysable bond (ester, peptide, etc.), which is formed via coupling to the hydrolysis of an energy-rich phosphate (P), which would draw the H_2O for hydrolysis from the carrier, on the cis side, as could be formulated as follows:

1. $R\big\langle{}^{OH}_{OH} + \sim P \rightleftharpoons R\!\!\bigcirc\!\!O + -P + H_2O$

$R\big\langle{}^{OH}_{OH}$ standing for Y, and $R\!\!\bigcirc\!\!O$ for X, respectively.

The after-reaction, taking place on the trans side, would then be

2. $R\!\!\bigcirc\!\!O + H_2O \rightleftharpoons R\big\langle{}^{OH}_{OH}$

Owing to strict spatial separation of the enzymes catalyzing the two steps, i.e., the main reaction takes place only on the left side of the barrier and the after-reaction, on the trans side, the overall reaction

$$\nu P \xrightarrow{\text{+H}_2\text{O}} P_i$$

can be considered *heterophasic*, since one of the reactants, here H_2O, must obligatorily come from the other phase.

We shall, however, keep the issue simple by maintaining that in this system, the primary transport reaction is crucial for the coupling at step 1 and hence scalar, whereas the translocation step and the reformation of Y are merely "uncoupled" secondary processes.

In order for the source-and-sink mechanism to become effective in active transport, the interconversion between X and Y must be associated with a change in certain properties of the carrier. The change may concern the affinity of the carrier for the transportee, or the velocity of the resulting carrier solute complex, or both. Accordingly, we speak of an *affinity* effect of the activation to the extent that active carrier (X) has a higher affinity for the transported solute (A) than has the inactive carrier. This effect alone, provided that translocation rate coefficients of all carrier species are equal, suffices to bring about active transport. On the other hand, we speak of a *velocity* effect to the extent that the complex of the active carrier with the transported solute (AX) translocates relatively faster than does the corresponding complex of the inactive carrier (AY). This effect by itself also brings about active transport, provided, however, that the rate coefficients of the empty carrier species meet certain requirements, which will be explained in more detail in the quantitative treatment farther below. Both affinity and velocity effects together should reinforce each other (HEINZ et al., 1972).

We may briefly try to obtain an intuitive picture of each of these effects. Owing to the coupling the concentration ratio of total active carrier to total inactive carrier speed will be higher on the cis side than on the trans side

$$\frac{X'}{Y'} > \frac{X''}{Y''} \tag{6.8}$$

To demonstrate a pure affinity effect we assume that X has a higher affinity for A than has Y. Consequently a gradient of AX down from cis to trans will form and cause net transport of A.

To demonstrate a pure velocity effect we assume that AX can penetrate the barrier while AY cannot. X and Y can also penetrate. It can easily be seen that AX (and X) will move from cis to trans, in the steady state only in exchange for the opposing movement of Y (and X) since AY cannot move.

The effects should be the same for primary and secondary active transport, provided that the coupling occurs by the source-and-sink principle.

Though both effects are energetically equivalent, kinetic differences are to be expected between the two: The affinity effect will vanish as the loading of the transport systems reaches saturation (K-type) whereas the velocity effect should be maximal under these conditions. We see that to the extent that the source and-sink system applies, primary and secondary transport have features in common. So the "push" type of primary active transport, in which the *activation* of the carrier (Y → X) is coupled to the chemical reaction, can be regarded as analogous to cotransport (or symport), whereas the "pull" type, in which the *inactivation* of the carrier (X → Y) is coupled to the driver reaction, would accordingly correspond to countertransport (antiport). For the principles of coupling just discussed it is also unimportant whether the translocation is due to a translatory movement, a rotation, a gate-type, or a conformational change of the carrier molecule.

6.1.6 Vectorial Coupling — The Conveyor Principle. In describing the source-and-sink mechanism, we have treated translocation (step 2) only as a passive movement of the complex between the transportee and the transport-active form of the carrier, XA, down its gradient. This gradient was said to be maintained by the activation-inactivation reactions of the carrier, one of which had to be directly coupled to a driver reaction. In other words, the translocation step was presented as not directly linked to a primary chemical reaction. In view of Curie's theorem (*v.s.*), which forbids, for anisotropic medium, the direct coupling of a (scalar) chemical reaction with a (vectorial) transport process, it was previously assumed that only a system of the source-and-sink type could possibly account for the linkage between two such processes (JARDETZKI and SNELL, 1960), since the vectorial step 2 or 4 (Fig. 14) could be directly coupled only to another vectorial process, such as in secondary active transport but not directly to a primary chemical reaction. This view has meanwhile been challanged, and it has been postulated that also a translocation, such as in step 2 and step 4, could be directly coupled to a chemical reaction. MITCHELL (1960) and KEDEM (1960) pointed out that a source-and-sink mechanism is not the only one thinkable for linking a chemical reaction with a transport process. So, according to these authors, a single enzymatic reaction might suffice to drive a macroscopically demonstrable transport process, provided that the system is "anisotropic", for instance, in that the enzymes concerned are spatially oriented within the membrane structure in a uniform way. Under these conditions, the (microscopic) movement of reactants (groups) of a chemical reaction, i.e., the "diffusion of chemical groups along or through the enzyme molecules" (MITCHELL, 1960) may be directly utilized for (macroscopic) translocation across the whole membrane. Based on Mitchell's ideas models can be devised that should afford active accumulation by a single chemical reaction without a mobile carrier. Such a mechanism would differ from the previously discussed source-and-sink mechanism in that the immediate force driving the solute across the barrier is not osmotic, i.e., not due to the concentration gradient of a carrier substrate complex, but is of the same nature as the affinity of a chemical reaction, here identical with the enzymatically controlled affinity between the transported solute and the acceptor molecule. It has to be pointed out, however, that such coupling has so far been demonstrated only with "group transfer", i.e., the transportee, which has to enter a primary chemical reaction, cannot do so as a solut

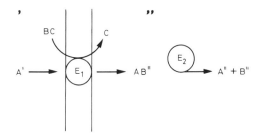

Fig. 16. Group translocation
system. For details see text

but only as a group, namely, by forming a covalent bond with
some other compound.

To visualize such a coupling we have to realize that the "common
step", i.e., the step shared between the transport and chemical
reaction is the translocation step (step 2). It is clear that
for this reason also the chemical reaction must have a vectorial
component, in other words, it must be, at least partly, "heterophasic". It is obvious that an enzymatic reaction can be heterophasic only if the catalyzing enzyme built into the barrier is
strictly oriented, as is already implicit in the postulate that
the membrane has to be "anisotropic" or, more specifically for
the direction of the transport, "asymmetric". The enzyme ought
to be oriented, for instance, so that it is accessible for only
one of the reactants from one phase but for only one of the products from the other phase. Under these circumstances only *one*
single enzyme would be crucially involved, in contrast to a
source-and-sink type model, although additional enzymes may be
present for auxiliary functions, as we shall see later in connection with the phosphotransferase system.

The enzyme immediately involved in the translocation would, in
effect, behave as if it were encased between two membranes of
different permselectivities in the following manner. Let us assume a reaction in which A reacts with BC to form the products
AB and C, according to the equation

$$A + BC \rightarrow AB + C \tag{6.9}$$

Of the two membranes that include the enzyme, the left one is
permeable to A but impermeable to AB, whereas the right membrane,
conversely, is permeable to AB but not to A. For the permeability
to BC and C, no restraints are necessary (Fig. 16).

The two membrane arrangement separates two bulk phases, ' and ".
We can predict that due to the coupled reaction A will be removed
from side ' and appear as AB on side ". To produce a transport-
like effect, a second enzyme, E_2, located in the right compartment is required to catalyze the transition of AB to A + B.

We see that A is translocated between the two bulk phases not as
a solute but, at least part of the way, as a chemical group of

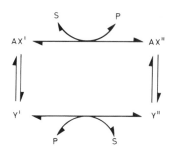

Fig. 17. "Conveyor" principle of coupling. The translocation of either the activated carrier (X) or the inactivated carrier (Y) is directly linked to a chemical reaction ($S \rightarrow P$)

AB. Hence this mechanism is called "group translocation". The translocation step itself is a primary chemical reaction, involving the dissolution and formation of (primary) covalent bonds. It is therefore no real transport process in the sense considered previously. However, if we include the second enzyme-catalyzed step, which is not vectorial, the overall effect amounts to true active transport of solute A, as by one of the previously discussed transport mechanisms. A more detailed discussion of the group translocation will be given later in this section in connection with the phosphotransferase system.

Another way to link the translocation step (step 2) directly to a chemical reaction has been suggested on the basis of a *conformational* model, in which translocation is postulated to be brought about by a conformational change of the carrier molecule (VIDAWER, 1966; EDWARDS, 1973; STEIN et al., 1973; KLINGENBERG et al., 1976). It has been surmised that this conformational change be energetically coupled to an exergonic chemical reaction ($S \rightleftharpoons P$), so that chemical energy liberated by the latter would be transduced into "conformational" energy, i.e., by shifting and maintaining the distribution ratio of the conformers at disequilibrium. The overall process would be represented by the following cycle (Fig. 17).

Y' and AX', on the one hand, and Y" and AX" on the other, represent the conformational states of the carrier. The prime-marked one exposes its binding site for A to the cis (') compartment, whereas the double prime-marked one exposes this site to the trans (") compartment. This kind of conformational transition involves a translocation of these sites and should not be confused with conformational changes that might be involved in the activation or inactivation of the carrier in the same position. The distinction between an active and an inactive form of the carrier, like that made in the system with scalar coupling, is not essential in the present system. Nevertheless, the notation X and Y will be maintained, but only to indicate that they are different and distinguishable for reasons given farther below. We are actually dealing with at least four different forms of the carrier in the present model, two of them exposed to the cis side (AX', Y') and two exposed to the trans side (AX", Y").

In order for active transport to occur, it is essential that the two vectorial transitions, $AX' \leftrightharpoons AX"$, $Y' \leftrightharpoons Y"$, be handled differently in coupling. Unless both transitions are coupled, each in a different direction, which seems very unlikely, it is necessary that only one is coupled, whereas the other must proceed spontaneously.

Again we have two separate pathways of the same transition, as in a source-and-sink system. In contrast to the latter, however, in the present model the two pathways are not separated by the location in the two different membrane faces, but by the selectivity of the coupling device: In other words, this device, possibly the enzyme catalyzing the coupling, must be capable of distinguishing between the loaded and unloaded state of the carrier: It must recognize the load. One might speculate that the attachment of A to the carrier induces a conformational change in the carrier ($Y \overset{A}{\to} AX$) by which the latter becomes reactive with the driver reaction

$$AX' + S \longrightarrow AX" + P.$$

After A is released on the trans side, the carrier resumes its Y conformation, which readily undergoes the uncoupled conformational change

$$Y" \longleftrightarrow Y'.$$

In such a system one could predict for the steady state that, in contrast to the source-and-sink mechanism,

$$\frac{ax'}{y'} < \frac{ax"}{y"}$$

As has been mentioned before, step 4, the restoration step of the transport cycle, could also in principle be coupled vectorially to the driver process. In that case it would be step 2 that should proceed spontaneously. Coupling at step 4 would, however, clearly be *indirect* with respect to the transported solute. Hence the ratio of loaded to unloaded carrier species should be higher on the left side than on the right side [Eq. (6.2)].

The models of conformational coupling may have some appeal but are highly speculative. Though it is widely believed that transitions of the postulated kind between a cis-oriented and trans-oriented state of the carrier may occur and give rise to a translocation of A, it is difficult to imagine how such a transition is coupled with a chemical reaction, and how such coupling could discriminate between loaded and unloaded carrier.

The conformational principle has also been applied to the model of gated channels: The channel-forming protein(s) are assumed to alternate concertedly between two asymmetric conformational states, one of which is open to the left side of the barrier only and the other to the right side only (KLINGENBERG et al., 1976). It would appear that such a model would function in principle like an asymmetric variant of Patlak's early gate-type model (PATLAK, 1957). The latter was shown by Patlak's calculations to be indistinguishable from a corresponding mobile-carrier model. This may also apply to the new asymmetric gate-channel model if the asymmetry is duly taken into consideration.

The conformational model in its present stage is so vague and so incompletely characterized that one cannot yet decide whether it is theoretically sound. So it has not been specified whether it works by group translocation, i.e., via covalent binding of the transportable solute to some acceptor, or whether the driver reaction has to be vectorial, e.g., in that S has to enter the reaction from one side of the barrier, and P is being delivered to the other.

6.2 Secondary Active Transport

6.2.1 Mechanism. As stated previously, coupling between two solute flows may lead to the transfer of energy between the two: One of the solute species, let us call it the species of the driven solute, may be transported against its electrochemical potential gradient provided that the flow of the other species, which would then be called the species of the driver solute, is downhill, and thus exergonic. To the extent that the gradient of the driver solute is continuously restored by an active process with the expenditure of metabolic energy, the uphill movement of the driven solute also becomes dependent on metabolism and thus meets the two criteria adopted by us to characterize active transport, namely the presence of a nonconjugate driving force and dependence on metabolism. Since such transport is not directly linked to a chemical reaction, it is called "secondary active". This name implies that in order to maintain transport the gradient of the driver solute has to be continuously restored, presumably by a "primary active" transport process. As to the mechanism of coupling in secondary active transport, any kind of interaction between the two solute species during their penetration, be it by friction, by direct association etc., could under suitable circumstances lead to the above-mentioned coupling phenomena. As has already been pointed out, however, in biological transport the term secondary active is usually meant to refer only to the stoichiometric coupling that is mediated by a "carrier"-like agent.

The coupling between the two solute flows could be *parallel* or *antiparallel*. In the first case, it is called *positive* and we speak of cotransport or symport. In the second case, the coupling is *negative* and we speak of countertransport or antiport. The observations that lead to the postulation of cotransport or countertransport in most cases are the following:

1. The phenomena of *cis* or *trans stimulation*, respectively. In other words, the transport of the nonelectrolyte is strongly stimulated by the addition of driver ion to the cis side in cotransport, or to the trans side in countertransport. From thermodynamic considerations, each effect should be reciprocal: The transport of the driver ion can also be stimulated by the addition of the appropriate nonelectrolyte solute to the cis or trans side, respectively.

2. In cotransport, the direction of the solute transport should be *congruent* to, i.e., the same as, the direction of the gradient of the driver ion. Accordingly, the transport should be reversed if the gradient of the driver is also inverted. For countertransport the oppositve should be true.

3. The movement of the driven solute should always be linked to that of the driver ion at a fixed stoichiometry, even though this stoichiometry may be obscured by leakages.

4. Even during complete metabolic inhibition, active transport of the driven solute should be demonstrated, as long as an appropriate gradient of the driver ion exists.

5. The power released by the downhill flow of the driver ion (input) should exceed the power required for the transport of the driven solute (output). In addition, the power available should actually be utilized for the transport of solute. This can be shown by measuring the degree of coupling, q, which will be discussed in Chapter 7.

The driver solute is usually an electrolyte ion, Na ion predominating in animal cells, and H ion in microorganisms. The active accumulation of many nonelectrolyte solutes, especially amino acids and sugars, is now believed to be driven by gradients of one of these ion species. The transport of sugars and amino acids appears to an example of cotransport (or symport). Also antiport of these substrates, e.g., with the efflux of K ions, has been considered, but has never been unequivocally verified. True antiport, however, appears to occur with anions, e.g., between Cl ions and HCO_3^- ions across the red blood cell membrane, between ATP and ADP, or between various organic anions across the inner mitochondrial membrane. In addition, it has been postulated that some amino acids may accumulate inside the cell predominantly by hetero-exchange, i.e., by exchange with another amino acid species, previously accumulated inside the cell. This kind of transport has also been called *tertiary* active, implying that the driving amino acid has previously been actively transported into the cell by symport with Na ions.

6.2.2 *Ion-linked Cotransport (Symport)*. The following models have been suggested for the mechanism of secondary transport. For cotransport it is assumed that the two solutes cotransported combine with the transport carrier at distinct sites, to form a ternary complex. The mere formation of such a complex, however, would not lead to energetic coupling unless it affects either the *affinity* between carrier and solute or the *velocities* of the resulting complexes, or both (HEINZ et al., 1972):

1. An "affinity" effect should result if the binding of the one solute raises the affinity of the carrier for the other solute, and vice versa. This effect by itself affords energetic coupling, even if the mobility of all carrier species were the same. Models based on this assumption have been called affinity-type models.

2. A "velocity" effect, on the other hand, should result if the ternary complex (and the empty carrier) penetrate the membrane faster than either of the two possible binary complexes. This effect by itself also affords energetic coupling, even in the

absence of the above affinity effects, provided that the rate constant of the ternary complex exceeds those of the binary complexes AX and BX. The velocity effect is enhanced if the mobility of the empty carrier exceeds those of the two binary complexes. Naturally, combination types are also thinkable, and the two kinds of effects, if present simultaneously, should reinforce each other.

To tell these two model types apart experimentally is difficult. A priori, one should expect certain characteristic differences in the kinetic behavior: In the pure affinity-type model, the presence of the driver ion should act by decreasing the apparent Michaelis constant. On the other hand, in the pure velocity-type model, the driver ion should increase the maximum transport rates A major complicating factor is the electrical potential, which is likely to be always present across the transporting barrier. Since the driver solutes are mostly ions, some of the resulting complexes with the carrier will be electrically charged and thus susceptible to the driving force exerted by an electric field. These effects may severely obscure the affinity and velocity effects, as will be discussed later in another context. Accordingly investigations of the behavior of the two Michaelis parameters during cotransport in various systems have so far led to conflicting results. In many cases, both parameters were found to be affected, but in some cases the maximum rate appears to be affected exclusively, and in other cases, exclusively the Michaeli constant. So at the present juncture, no system has been successfully identified in this respect. It could also be that the mixed type is present in all systems, but that, depending on certain circumstances, the one or the other Michaelis parameter may suffer a greater change in the presence of the cosubstrate.

Analogous models may be developed for coupling by countertransport. A kind of *affinity* effect would come about, if the binding of the driver solute reduced the affinity for the other solute, and vice versa. This effect would be strongest if either of the two antiporting solutes were bound at the same site, as in true hetero-exchange; however, binding at different sites if they affect each other, might also do. Also a kind of *velocity* effect could be construed for countertransport. For that purpose, the velocities of the two binary complexes, i.e., between the carrier and only one of the two solute species, would have to be much greater than those of the ternary complex and of the empty carrier. The quantitative relationship between the velocities is thus different from the corresponding one in cotransport, as will be shown later in the quantitative treatment. Since the movements of the coupled flows are antiparallel, the coupling between them is by definition *negative*. Secondary active transport appears to be of great significance in many biological transport systems and therefore deserves special consideration here.

6.3 Primary Active Transport by Solute Translocation

6.3.1 General Mechanism. In certain systems the uphill transport of a solute appears to be driven directly by the affinity of a chemical reaction, e.g., of the hydrolysis of energy-rich phosphates, without detectable participation of another co- or countertransport solute. It has been postulated that in such systems the translocation of the transportee is directly coupled to the "advancement" of the chemical reaction. This kind of coupling, which one could appropriately call chemi-osmotic, even though it can be easily handled in terms of thermodynamic formalism, poses some fundamental problems as far as a plausible mechanistic model is to be devised. For instance, the direct coupling between a chemical reaction (a scalar) and an osmotic process (a vector) is in seeming contradiction to Curie's theorem, which prohibits the direct coupling between processes of different tensorial order, as here between vector and scalar. This problem, however, is no longer considered a serious one, because Curie's theorem strictly applies only to an isotropic, symmetric system (KEDEM, 1960). By introducing a certain kind of asymmetry into the system, e.g., by choosing a membrane that does not allow the same reaction to take place on each side, the membrane can assume the function of a vectorial operator. Meanwhile many reasonable and probable models have been devised to allow for chemi-osmotic coupling in primary active transport.

In primary active transport we may distinguish between *solute translocation* and *group translocation*. In solute translocation the solute concerned is assumed to be transported as such, as in secondary active transport, namely, loosely bound to the carrier by a spontaneously dissociable, noncovalent bond. In group translocation, on the other hand, the solute must first undergo a chemical reaction with the carrier or with another compound involved in the transport system, thus passing the barrier as a covalently bound "group". This group may later be split off again, but not necessarily immediately after the translocation.

Most models of primary active transport with *solute translocation* are based on the source-and-sink principle, which we have characterized as involving two different chemical reactions that are spatially separated by the osmotic barrier. In the one reaction an active carrier is being activated and in the other, inactivated, e.g., by chemical or conformational changes. The active carrier produced in the "source" reaction would be able to combine with the transported solute, carry it across the barrier, presumably in the same way as in facilitated diffusion, and, after being inactivated by the "sink" reaction, delivers the solute to the other (trans) compartment. These two processes are envisaged to occur cyclically, in that the carrier in the inactivated state returns to the original interphase where it is activated again. Primary active transport systems of the source-and-sink type, like the secondary active transport systems, can thus be regarded as a special case of carrier-mediated (facilitated) diffusion, modified through coupling to an exergonic chemical reaction. Accordingly, this kind of active transport has many features in common with facilitated diffusion, such as saturation kinetics, competitive inhibition by other solutes competing for the same transport systems, and specificity.

In order for this transport to become active, either the activation or the inactivation step has to be energetically coupled to an exergonic process.

The source-and-sink principle has formed the explicit or implicit basis of numerous hypothesis of active solute transport. Yet it has hardly ever been unequivocally proven to exist anywhere in biological transport. In most cases it is invoked for transport systems that cannot be shown either to be secondary active, i.e., driven by an ion gradient, or to function by a group transloca- tion mechanism. Among these are, besides some unexplained micro- bial systems for sugar and amino acid transport, mainly ion trans port mechanisms, e.g., for the transport of Na^+, K^+, Ca^{2+}, Cl^-, etc. Some of these systems have been shown to be related to spe- cific "transport ATPases", i.e., ATP-splitting enzymatic activi- ties demonstrable in the membranes of cells and tissues contain- ing ion pumps. These ATPases are believed to be involved in the direct coupling of the hydrolysis of energy-rich phosphate bonds to the translocation of the solutes concerned across a biological membrane. One characteristic feature of these ATPases, which are firmly attached to, or built in the membrane, is that they are activated by the ion species whose transport they are supposed to promote. Whether there are also nonelectrolyte transport sys- tems that involve specific ATPases is still dubious and probably unlikely. At least, no ATPase has unequivocally been demonstrated to be activated by such nonelectrolytes as sugars or amino acids.

Typical examples of such ATPases are:

1. The Na^+, K^+ ATPase, which is activated by the joint presence of these two alkali ions, and which is supposedly involved in Na^+ and K^+ transport across the membranes of many cells.

2. The Ca^{2+}-activated ATPase, shown to be involved in the trans- port of Ca^{2+}, for instance in the sacroplasmic reticulum of mus- cle fibers and in the membranes of other cells.

3. The more recently discussed K^+-activated ATPase, considered to be involved in an active K^+-H^+ exchange, in the acid secreting cells in the gastric mucosa (Forte, Sachs).

4. Other ATPases, such as the Ca^{2+}, Ma^{2+}-activated ATPase of mitochondria, assumed to reversibly link the H^+ movement to ATP- hydrolysis or synthesis, respectively, presumably function by vectorial group translocation, and will be discussed later in connection with this kind of transport.

All of these ATPases require the presence of magnesium for their action.

These various types of ATPases also differ from each other with respect to their specific inhibitors. So the Na^+, K^+ ATPase is strongly inhibited by cardiac glycosides, especially by strophan- thin (ouabain), whereas the Ca^{2+}-activated ATPase is inhibited by ruthenium red.

6.3.2 Ca^{2+}-activated ATPase. The contraction of actomyosin in muscle requires the presence of Ca^{2+} in the cytoplasm of the muscle cells (HASSELBACH, 1974; MARTONOSI, 1972). Relaxation is, con- versely, effected by a rapid removal of Ca^{2+} by a Ca^{2+}-pump into the sarcoplasmic reticulum (SPR). This active transport is closel and reversibly associated with the hydrolysis of ATP: If the Ca^{2+} flow is forced into the reverse direction, i.e., out of the SPR

space, ATP may be resynthesized (HASSELBACH). The SPR material
has enzymatic ATPase activity that is activated by Ca^{2+}, and is
therefore assumed to be involved in the transport process. It
has also been shown that during the transport process a protein
component of the SPR, presumably participating in the transport
process, is temporarily phosphorylated by ATP, and looses the
P group after the Ca^{2+} has been released to the inside of the
SPR. Available evidence about the active Ca^{2+} transport in the
SPR of muscle cells could be fitted to a source-and-sink type
model. The Ca ions appear to be translocated after being bound
outside to a high-affinity form of the carrier, and released inside after conversion of the carrier to a low-affinity form.
The interconversion between the two forms of the carrier seems
closely related to phosphorylation of the protein, presumed to
be involved in the binding and translocation of the Ca^{2+}. Whether
the overall process is as simple as that the phosphorylated and
dephosphorylated protein, or vice versa, represent the active
and inactive forms of the carrier, respectively, in terms of our
general model, has to await further elucidation. Active Ca^{2+}
transport and an associated Ca^{2+}-activated ATPase have also been
detected in red blood cells (SCHATZMANN, 1966).

6.3.3 Na^+-K^+-activated ATPase.

In contrast to the Ca^{2+} pump, other
ion pumps have been shown to be exchange pumps, i.e. the active
transport of one ion species is forcibly linked to the active
transport of another ion species in the opposite direction. If
these species have the same sign and are exchanged stoichiometrically in electrically equivalent numbers, such a pump is "electrically silent". In some such pumps, however, the stoichiometry
is different, so that electrical net charges are moved in a given
direction. Such pumps act like generators of electrical current
and are therefore called "electrogenic".

We have already seen that exchange pumps can be envisaged in
terms of the source-and-sink type of coupling. One possibility
would be to link two source-and-sink systems in antiparallel directions. As we show below (p. 113), a *single* source-and-sink system can, at least theoretically, be fitted to perform active exchange transport if one merely assumes that the postulated inter-
conversions between the two carrier forms are accompanied by
drastic changes in affinity. A typical example and one of the
most extensively studied systems of an (electrogenic) exchange
pump is the Na^+-K^+ pump, present in many living cells. Its ener-
gization is assumed to be achieved by the hydrolysis of ATP via
the well-known Na^+-K^+-activated ATPase. This assumption is based
on a close correlation between this enzymatic activity and Na^+-
K^+ pump activity, as seen from the following observations.

1. Both the Na^+-K^+ pump and ATPase require, for maximum activity,
the simultaneous presence of Na^+ and K^+, each in about similar
concentrations.

2. Both pump and ATPase are inhibited by cardiac glycosides added
in similar concentrations.

3. Both pump and ATPase are located in the membrane.

4. The activity of the pump and that of the ATPase in a given
membrane are closely related to each other: The stronger the

pumping activity of a given membrane, the more ATPase activity is present in this membrane.

5. In some cases it could even be shown that transport turnover per pumping site is of a similar order of magnitude as the rate of ATP-splitting per ATPase site of that particular membrane tissue.

Alternative hypotheses concerning the mechanism by which the translocation of the ions is tied to the chemical hydrolysis conflict with each other. The main discrepancy between these hypotheses concerns the question of whether the two ions are translocated *sequentially*, one after the other in two discrete steps, or whether they are translocated precisely *simultaneously*, in a single step. As a third possibility, it has also been suggested that Na^+ and K^+ are fundamentally transported sequentially, but that the two steps overlap to such an extent that the behavior of the mechanism comes close to the strict simultaneous mechanism. Another point of argument between the above two hypotheses concerns the so-called phosphorylated intermediate. It has been clearly established that the terminal phosphate of an ATP molecule before it is liberated is transiently covalently bound to the enzyme. There is little doubt that this "phosphoenzyme" really appears during these reactions; it can be most easily demonstrated by the appearance of ^{32}P activity in the enzyme protein after the addition of ^{32}P-labeled ATP or, under special conditions, after the addition of labeled inorganic phosphate (P_i). The sequential hypothesis postulates that this intermediate has an essential function within the linkage between ATP splitting and ion transport. The simultaneous hypothesis, on the other hand, doubts this and considers the phosphorylated intermediate either as an artifact or at least attributes its formation to a side path not essential for the main reaction. It seems clear that the phosphorylated intermediate contains the phosphate group as an acyl phosphate of a specific aspartyl site chain of the enzyme.

An elaborate scheme has been developed to explain the various events in terms of the *sequential hypothesis*. We postulated that the enzymes occur in two conformational states, E_1 and E_2. Na^+ binds preferentially to E_1. In the presence of ATP and Na^+ the enzyme becomes phosphorylated to form $Na^+ \cdot E_1 \sim P$. This intermediate is still a "high-energy" compound, and accordingly, the phosphorylation is fully reversible; in the presence of excess ADP, the enzyme is dephosphorylated and ATP is resynthesized. If labeled ADP is added, the same label can be shown later in the chemical ATP, a phenomenon called ATP-ADP exchange (POST et al., 1960, 1973).

The phosphorylation reaction is believed to be coupled to the movement of Na^+ across the membrane. If the Na^+ is removed from $E_1 \sim P$, the latter appears to undergo a conformational change to form $E_2 - P$. $E_2 - P$ is a "low-energy" phosphate compound, in other words, through the conformational change the free energy of hydrolysis is greatly reduced. Accordingly, the transition from $E_1 \sim P$ to $E_2 - P$ is irreversible, except under very special conditions, to be discussed later. Furthermore, no ATP-ADP exchange is possible in the presence of $E_2 - P$ only. It is essential for the overall reaction that the transformation $E_1 - P$ to $E_2 - P$ also effects a shift in affinity: $E_2 - P$ strongly prefers K^+ to Na^+. In the presence of K ions, $E_2 - P$ rapidly forms a complex with K^+, which is then translocated to the left side where it looses its phosphate group, and later the K ion. Since the movement of K ions across the membrane into the cell is believed to be a prerequisite for these last reaction steps, the final dephosphorylation takes place inside the cell. This dephosphorylation is to some extent reversible, which is plausible in view of the low energy content of the P group in $E_2 - P$. Provided that the level of inorganic P is sufficiently high, the K^+-binding enzyme can be rephosphorylated, which can be shown by the use of isotopically labeled inorganic phosphate. Na ions counteract this rephosphorylation so that with the high Na^+ concentration usually present in the experimental systems, the reversibility of the P splitting of $E_2 - P$ does not show. Instead, under these conditions, the Na ions will replace the K ions with the enzyme and thereby effect the reconversion of the enzyme into conformation 1, so that the ATP-splitting and ion-transporting cycle can begin anew.

This hypothesis implies that the ATPase molecule functions alternatingly as a carrier for Na^+ and K^+, respectively. It is supposed to bind the Na^+ and K^+ at the same site, Na ions in the E_1 form and K ions in the E_2 form. The phosphoryl group is also assumed to be bound to the same site, whether it comes from ATP, in the E_1 conformation, or whether it comes from inorganic phosphates, in the E_2 conformation. The cardiac glycosides, such as ouabain, are believed to combine preferentially with $E_2 - P$, thus preventing its reconversion to the E_1 conformation.

The simultaneous hypothesis (CHIPPERFIELD and WHITTAM, 1974), on the other hand, postulates that Na and K are bound simultaneously to different and specific sites of the enzyme, Na to an inwardly directed site and K to an outwardly directed site. In the presence of ATP and magnesium, ATP is transiently bound to the enzyme and immediately hydrolyzed. This hydrolyzation is associated with a conformational change of the enzyme whereby the two ion-binding sites are shifted in space, so that Na is carried to the outside, and K to the inside interphase of the membrane. Subsequently, the enzyme returns to its original conformation and the process can start anew. No stable phosphorylated intermediate is formed during this process, except for a very transient complex between the enzyme and the energy rich phosphate, which is unstable and cannot be demonstrated. The phosphorylated intermediate is formed only if the main reaction is prevented from proceeding normally. The main evidence in favor of this simultaneous model is derived from kinetic observations.

Both hypotheses have their enzymatic counterpart. For example, a group can be transferred from one compound to another enzymatically either in a one-step or in a two-step reaction. In the first case, both compounds combine with the enzyme so that the transfer can take place, and afterward the two products leave the enzyme. In the two-step reaction, the enzyme first combines with one substrate, then binds the transferable group, releases the first product, and then combines with the second substrate

to which the group is transferred. Kinetically the two mechanisms
are distinctly different from each other. The kinetics of the
transport ATPase reaction seems to fit that of the one-step reaction better than that of the two-step reaction. If, however,
as has already been mentioned, they are overlapping in the sequential reaction between the two steps, Na extrusion and K uptake, the sequential reaction may become kinetically indistinguishable from a simultaneous reaction. Further research appears
to be necessary to solve this problem.

The "sequential" hypothesis could easily fit the combined source-
and-sink model discussed above. The assumption of a high-energy
phospho-enzyme $(E_i \sim P)$, which would take the function of our X',
and of a low-energy P-enzyme, taking the function of our Y',
makes it somewhat more complicated because we obtain a total of
three different states of the carrier, if we include the unphos-
phorylated enzyme as an additional stage. However, it is still
undecided whether the phospho-enzyme found is an essential intermediate of the ATPase reaction rather than an artificial side
reaction. Crucial is the assumption that the two alternating
transitions between X and Y, both the energetically coupled one
and the spontaneous one, are associated with a rather drastic
change in affinity for the two solutes A and B, which could stand
for Na^+ and K^+, respectively.

The "simultaneous" ATPase hypothesis, on the other hand, seems
difficult to reconcile with a source-and-sink system. The translocation step is assumed here to involve a conformational change
associated with a shift in position of the binding sites: Conformation X', in which the Na^+-binding site is exposed to the
inner face of the membrane, and the K^+-binding site to the outer
face, and a conformation X", in which the positions of the binding
sites are conversed. Active exchange could best be fitted to a
modified conformational model, which has been discussed in connection with vectorial coupling. In the present system one would
have to postulate direct coupling between the hydrolysis of ATP,
and either the conformational change of the carrier in the loaded
form left to right (step 2) or the corresponding conformational
change of the unloaded carrier in the reverse direction. Again
the fundamental question arises as to whether such a coupling,
which implies the transduction of chemical energy into "conformational" energy, is possible. This question cannot be answered
at the present time.

6.4   Primary Active Transport by Group Translocation

6.4.1 General Mechanism. There is evidence, or at least a plausible
basis, for the assumption that some solutes, in order to be translocated across a barrier, have to undergo a chemical reaction
that links them by a covalent bond to the carrier or to another
compound. They are thus carried through the barrier as a chemical
"group" rather than as a solute. Such group-translocating systems
can be looked upon as a special kind of enzymatic transfer reaction, by which a reactive group is transferred from a donator to
an acceptor, e.g., the transfer of an acetyl group via coenzyme
A to choline. The group-translocating system differs from these

chemical reactions in that the transfer of the group is associated with the simultaneous translocation of this group through a barrier.

The term group translocation is often used to imply the presence of vectorial coupling. There may be no cases of vectorial coupling other than with group-translocating systems. The converse, however, is not necessarily true. We can at least visualize certain group translocation processes proceeding by scalar coupling, analogous to scalar coupling in solute-translocating systems. Let us, for instance, assume that the transportee is being bound covalently to a carrier, perhaps by a coupled reaction, subsequently moved through a barrier as a "group", and finally released on the trans side again as the free solute. Such a system can easily be described in terms of a source-and-sink mechanism, because within the overall transport process a clear distinction between chemical (binding and release) reactions, and translocation, can be made; in other words, the binding and/or release reaction could be the one that is coupled directly to the driver reactions, whereas the translocation step is merely passive and dissipative, i.e. XA, the group-carrying translocator moves down its electrochemical potential gradient, similarly to the previously described mechanism of solute translocation. In analogy to the corresponding solute-translocating system either the formation of this bond on the cis side (push effect) or the dissolution of this bond on the trans side (pull effect) would have to be coupled. The first coupling would presumably require the prior activation of one of the reactants, i.e., the carrier moiety (X) or the substrate moiety (A), as is depicted in Figure 18.

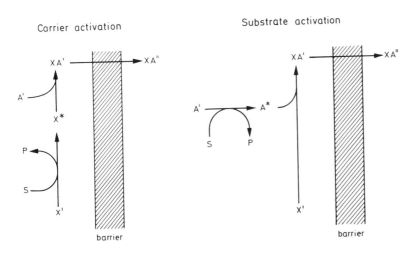

Fig. 18. Group translocation via source-and-sink system. The solute A forms a covalent bond with the translocator X. The energy for this bond formation is provided by prior activation of the carrier, or of the transported solute. For details see text

6.4.2 The Carnitin Cycle, by which fatty acid groups are translocated through the mitochondrial membrane, could serve as an example of the substrate-activated group translocations: The fatty acid (F) is first activated by ATP and coenzyme A (CoA-SH) and then transferred to carnitin to form fatty acyl carnitin (F-Car). This penetrates the membrane and subsequently transfers the fatty acyl group to another CoA-residue:

$$F' + ATP' + CoA-SH' \rightleftharpoons EoA-S-F' + ADP' + P'_i$$

$$CoA-S-F' + Car' \rightleftharpoons F-Car' + CoA-SH'$$

$$F-Car' \rightleftharpoons F-Car''$$

$$F-Car'' + CoA-SH'' \rightleftharpoons Car'' + CoA-S-F''$$

$$Car'' \rightleftharpoons Car'$$

This transport would obviously become active if the fatty acid after the translocation would merely be released hydrolytically rather than transferred to CoA. According to this model the system might well work according to the source-and-sink principle, since the translocation, i.e., the penetration of the membrane by the fatty acyl carnitin compound, can be clearly visualized as an osmotic step distinct from the chemical binding and release reactions.

Another example is the redox pump of protons, which in contrast to the proton translocation ATPase could also be treated in terms of the source-and-sink system:

6.4.3 The Redox Proton Pump. The redox pump as a hypothetical model of H^+-transport is very old and was, for instance, invoked for gastric acid secretion several decades ago. Later it was dropped because of findings supposedly indicating the direct involvement of ATP in gastric and acid formation. At that time the redox pump and generation of protons by ATP were considered mutually exclusive, a view which can no longer be maintained. For microorganisms and mitochondria, the evidence in favor of a redox pump is strong and is supported by the occurrence of the appropriate oxido-reductases and redox carriers within the membranes of these organisms. In animal cells, however, evidence that their plasma membranes contain such compounds is still scanty and dubious. The observations with microorganisms and mitochondria can therefore not simply be extended to the plasma membranes of other cells, e.g., animal cells, even though sizable proton secretion has been suspected there. Whether special acid-secreting cells, such as the parietal cells in the gastric mucosa or the corresponding cells of the kidney represent an exception, is not yet known.

The older view of the redox pump was based on the model of the "electron shuttle". The crucial reaction was considered the exchange of electrons between a hydrogen donor and an electron acceptor, for instance between a flavoprotein (FH_2) and a heavy metal-ion complex such as cytochrome (Cyt):

$$FH_2 + 2\ Cyt^{+++} \rightleftharpoons F + 2\ Cyt^{++} + 2\ H^+$$

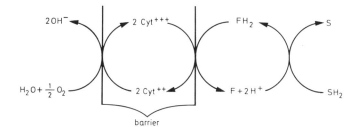

Fig. 19. Redox pump for protons via electron shuttle. Hydrogen donnor (SH_2) reacts from the right side and the electron acceptor (oxygen) reacts from the left side. Only electrons are moved across the barrier. Protons are accumulated on the *right side* and OH ions on the *left side*

Depending on the direction, this reaction liberates or consumes 2 H^+, respectively; if FH_2 is continuously regenerated by a reducing substrate (SH_2), and if Cyt^{2+} is continuously reoxidized, ultimately by O_2, a steady generation of H^+ and OH^-, according to the overall reaction,

$$SH_2 + 1/2\ O_2 + H_2O \longrightarrow S + 2\ H^+ + 2\ OH^-$$

can be maintained, provided that the combination of H^+ and OH^- is prevented by an appropriate device. The simplest model would be an electron-conducting membrane that allows reaction with the hydrogen donor (FH_2) only at one interface and reaction with an electron acceptor, e.g., O_2, only at the other, and that is impermeable to H^+ and OH^-. The electron conduction could be achieved by a mobile cytochrome-like carrier within the membrane (Fig. 19). Obviously, this model, which was formerly thought to underlie all H^+-transport systems, implies that substrate-H_2 is the only source of H^+. This would strictly limit the maximum H^+/O_2 ration to 4.

This model clearly does not transport H^+ or OH^- but only electrons, which under appropriate conditions may lead to the accumulation of H^+ and OH^- ions in the two compartments. A true H^+ pump will be described in Chapter 7.

In the newer version of the redox pump the limitation of the H^+/O_2 ratio to the value of 4 is no longer required. MITCHELL (1961) pointed out that many more protons per O_2 could, at least theoretically, be produced, depending on the number of loops, or on the number of reaction steps in series of the type described above (Fig. 20). A model can easily be construed in which a hydrogen shuttle replaces the electron shuttle mentioned above. On the cis side, a pair of protons are picked up by a redox reaction, and on the trans side, the protons are liberated by a similar reaction. In other words, a sequence of electron carriers alternating with hydrogen carriers would be required to be arranged in the membrane in such a way that hydrogen atoms are transported only in one direction, and electrons only in the other, as illustrated in Figure 21. In such a model the number of protons generated per O_2 molecule is limited only by the number of alternating sequences (loops), and by the energy available.

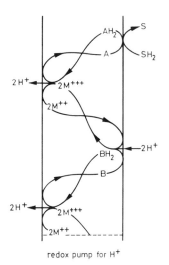

redox pump for H⁺

Fig. 20. Redox pump for H^+. The reaction pathway of the redox steps are represented by
loops. The mechanism can be extended by further loops at the lower end. The essential
factor is that a hydrogen-transporting molecule alternates with an electron carrier. The
figure is a modified representation of
Mitchell's hydrogenation/dehydrogenation
mechanism

Since this redox pump transports protons not as such but as part
of an organic molecule, namely, as an H_2 group, it is basically
a mechanism of *group translocation*. Obviously most of the protons
derive here from H_2O. As mentioned at the outset of this chapter, this kind of group translocation is not much different from
previously described systems of primary solute translocation.
H^+ ions are picked up on the cis side and as such released on
the trans side. Hence the system can be treated as a source-and-
sink system, at least if it functions as usually visualized, and
as may be illustrated at a single loop of the redox chain depicted above. H^+ ions from the cis side are translocated as H-
groups by a hydrogen carrier, for example, a quinone derivative
(CoQ). The assumed mechanism would consist of the following four
steps in terms of the above model:

1. $2H^{+\prime} + CoQ' + 2(Me_1^{2+})' \longleftrightarrow CoQ\ H_2' + 2(Me_1^{3+})'$
2. $CoQ\ H_2' \longleftrightarrow CoQ\ H_2''$
3. $CoQ\ H_2'' + 2(Me_2^{3+})'' \longleftrightarrow CoQ'' + 2H^{+\prime\prime} + 2(Me_1^{2+})''$
4. $CoQ'' \longleftrightarrow CoQ'$

Translocation of H^+ occurs if Q reacts with Me_1' only on the cis
side, and with Me_2'' only on the trans side of the barrier. The
overall reaction

$$2\ Me_1^{2+} + 2\ Me_2^{3+} + 2\ H^{+\prime} \longleftrightarrow 2\ Me_1^{3+} + 2\ Me_2^{2+} + 2\ H^{+\prime\prime}$$

can take place only with the mediation of the hydrogen carrier
CoQ, as for instance via the cycle shown in Figure 21.

The model, at least, appears to distinguish between two spatially
separated scalar redox reactions and the vectorial translocation
of a hydrogen carrier. Such a system may be classified as a bor-

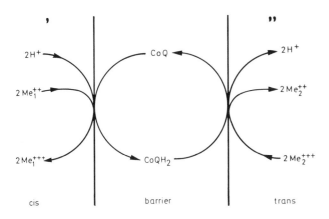

Fig. 21. Proton pump via H$_2$ shuttle. Hydrogen is shuttled across the membrane by CoQ. -H$_2$ is being formed from protons and electrons on the *left side* and delivered as protons on the *right side*. Electron donors (*Me$_1$*) react from the *left side* and electron acceptors (*Me$_2$*) react from the *rigth side*

derline case, e.g., as a group-translocating system with scalar coupling, since the vectorial translocation step is presumably driven by osmotic forces only, namely, the chemical potential gradients of CoQH$_2$ and CoQ. If compared with the carnitin system, the covalent binding of 2H$^+$ could in the redox system be characterized as carrier-activated: Only through reduction by Me$_1$ to CoQ^{2-} is Co$_2$ rendered capable of binding H$^+$ to form CoQH$_2$.

In the above mentioned cases, to the extent that the source-and-sink treatment can be applied, we could, regardless of what the detailed mechanism of coupling may be, operationally make a distinction between *affinity* and *velocity* effects, in analogy to those discussed previously in connection with solute translocation. This may be demonstrated for the redox proton pump. The overall transport rate, assumed to be determined by the translocation step, can be thought to be accelerated in two ways: firstly, by increasing the formation of CoQH$_2$, and secondly, by increasing the rate coefficient of its translocation. In the reaction between electron donor (Me^{2+}), H acceptor CoQ, and H$^+$, not only CoQH$_2$ but presumably to some extent also CoQ^{2-} or CoQH$_2^{2+}$ will be formed. If the latter two charged intermediates are, however, negligible as compared to CoQH$_2$, there would be only an affinity effect. If, on the other hand, the above ions form just as readily as CoQH$_2$, we could speak of a velocity effect. This latter effect is usually not considered a real possibility.

Redox mechanisms have also been suggested for the transport of solutes other than protons, i.e., for the active uptake of anions in plant roots, "anion respiration" (Lundegaard), or for the transport of Na and K ions in animal cell membranes (Conway). These hypotheses have been mostly abandoned, at least for the alkali ion transport in cellular membranes, partly for the already mentioned absence of suitable redox catalysts in the membrane, and partly because the direct linkage of these ion transports to ATP hydrolysis via the transport ATPase appears obvious.

6.4.4 The Proton-Translocating ATPase. A fundamentally different mechanism is attributed to another H$^+$-transporting system: The proton-

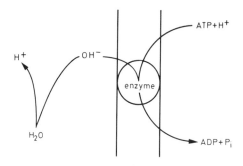

<u>Fig. 22.</u> Proton translocating ATPase. Schematic representation of reaction steps. The H_2O required for the hydrolysis of ATP is provided by OH^- from the *left side*, and by H^+ from the *right side* via a hetero-phasic reaction, catalyzed by a membrane-bound enzyme. For details see text

translocating ATPase. Here the coupling is vectorial, i.e., the *translocation* step is assumed to be immediately coupled to a chemical reaction (MITCHELL and MOYLE, 1958). The hydrolysis of the P-O-P linkage of the energy-rich phosphate is presumably hetero-phasic: The OH^- required for the hydrolysis has to be supplied from one phase only, in Figure 22 from the left side of the barrier.

As discussed before, we can represent the mechanism of this process simply by a model system in which the ATPase enzyme is encased between two membranes with different permselectivities. For example, the left membrane is impermeable to H^+, and the right one, impermeable to OH^-, whereas both membranes are impermeable to H_2O but may be permeable to ATP and its split products. If this chamber contains an active ATPase it can be predicted that the hydrolysis of each ATP should lead to a separation of H^+ from OH^-, virtually to an actively driven transfer of OH^- from left to right. We say "virtually" because the transport is not real, since the OH^- removed from the left side does not appear on the right side as such but as an OH-group of the ATP split products. This system could under suitable conditions act as an ATP-driven H^+ pump. In reality, however, it is neither a H^+ nor an OH^- transport system but a group-translocating system, in principle similar to that of the phosphotransferase system, which will be discussed subsequently.

6.4.5 The Phosphotransferase systems are perhaps the most typical and best elucidated examples of vectorial group translocation. Several microbial transport systems have been found to use such a vectorial group transfer system, e.g., for the transport of glucose and α-galactoside in various bacteria, *E. coli*, *S. typhimurium*, *Aerobacter aerogenes*, gram-negative bacteria, and also with membrane preparations derived from *E. coli* (KABACK, 1968). In these transport systems the accumulation of the mentioned substrate depends on the action of three enzymes: E. I, E. IIB, and E. III (or E.IIA) and a histidine-containing, low-molecular-weight protein (HPr) (POSTMA and ROSEMAN, 1976). However, only one of them (E IIB) is built into the membrane and appears to be directly concerned with the translocation as such, whereas the others have subsidiary, though indispensible functions. The overall process appears to involve the following steps. At the outside face of the membrane the substrate (sugar) is bound by enzyme IIB. Inside the cell the histidine-containing protein

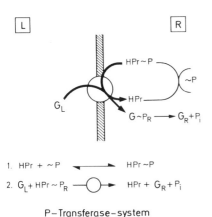

L   R

HPr~P

~P

HPr

G_L

$G{\sim}P_R \longrightarrow G_R + P_i$

1. HPr + ~P $\longleftarrow$ HPr~P

2. G_L + HPr~P_R $\longrightarrow$ ◯ $\longrightarrow$ HPr + G_R + P_i

P—Transferase—system

Fig. 23. Phosphotransferase system. Schematic diagram. G stands for glucose, with subscripts L and R for *left* and *right side*, respectively. *HPr* stands for histidine protein, in the phosphorylated form (*HPr* + $\sim$ P) and dephosphorylated form (*HPr*). P_i stands for inorganic phosphate

(HP_r) is phosphorylated by the energy-rich phosphate phosphoenol-pyruvate (PEP) with the mediation of enzyme I. The crucial step now is the transfer of this phosphoryl group to the sugar moiety bound to F IIB with the mediation of another enzyme (E IIA, or E III), and the subsequent release of the product, sugar-6-phosphate (G-6-P) to the cytoplasm. As a final step, G-6-P may be hydrolyzed there by a phosphatase, which may then lead to a true accumulation of free sugar in the cell. Using double labeling techniques, Kaback showed that only extracellular glucose is phosphorylated in reaction II, although its reactions partner, HPr $\sim$ P, and the resulting product, G-6-P, are only found inside the cell. If the preparations or cells were preloaded with [^{3}H] glucose and then resuspended in [^{14}C] glucose, the accumulated glucose 6-phosphate had exclusively the same label as the outside substrate (KABACK, 1968). Obviously in this system the membrane-bound enzyme II catalyzes a reaction between intracellular HPr $\sim$ P and extracellular glucose, across the osmotic barrier, by which the glucose appears to be "pulled" into the cell (Fig. 24).

It is often asked whether group translocation is true active transport. Certainly it does not fit the rigid definition given initially for transport processes as distinct from chemical reactions, since the primary result of group translocation is the appearance on the trans side of a substrate different from that removed from the cis side. So, for instance, in the phototranspherase system a sugar is taken up from the medium, but what appears first inside the cell is sugar phosphate. Likewise the OH$^-$ that is removed from the cis side by the H$^+$-translocating ATPase is delivered on the trans side as an OH group. A second process is required to transform the translocated group back into the original solute, such as an intracellular sugar phosphatase. In this way, a true accumulation of the sugar inside the cell may eventually be achieved, not by the translocation process itself but by a secondary reaction, which is not an integral part of the translocating mechanism. Hence we may look upon a group-translocating system, as upon any heterophasic reaction or as upon a hybrid between a purely chemical reaction and a transport

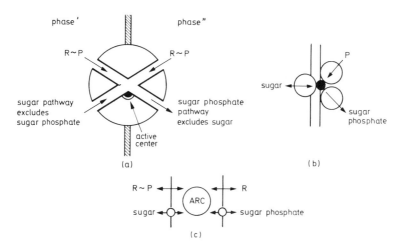

Fig. 24. Hypothetical mechanism of phosphotransferase system. In the figures the two mentioned models are schematically depicted. In (a) the active reaction center is assumed to be in the middle of the unit and accessible to the sugar and to the sugar phosphate, respectively, through specific channels. The two additional channels to the center are for the phosphoryl donor and are to indicate that this could approach the center from either side of the barrier. The model (b) refers to the same mechanism though here the two substrates, sugar and sugar phosphate, reach the center each by its specific hemicarrier. Each possible model could be functionally represented by a reaction center (c) secluded in a compartment by two different permselective membranes, the one transmitting the sugar but excluding the sugar phosphate and the other transmitting the sugar phosphate but excluding the sugar. The phosphoryl donor could deliver the phosphoryl group through either membrane; at least no exclusion is required for the proper function of the mechanism

process; the former having properties in common with both, i.e., the translocation of matter between two phases with a transport system and the alteration of this matter with a chemical reaction.

In accordance with the requirements of vectorial group transfer, as discussed before, it must be postulated for the PTS that the sugar group is admitted only from the *outside* of the cell whereas the sugar phosphate can be released only on the *inside* of the cell. A priori it is unimportant from which phase the phosphoryl group is supplied, although de facto it seems to come from the inside of the cell (KABACK, 1968). Hence the two pathways by which sugar and sugar phosphate, respectively, have access to the active center, must be specific, each accepting its own substrate and excluding others. Whether these pathways are selective channels within a single enzyme unit (Fig. 24) or whether they refer to different mobile hemicarriers, each of which transports its specific substrate to the reaction center but not beyond, is not known, nor are other mechanic details of the system known.

Anyway, a clear distinction between chemical reaction and trans-location, here between the phosphorylation of the sugar and the translocation of the glucosyl group across the barrier, is not possible: Both are different phenomena of one and the same pro-cess. One should expect that in such a system the translocation stops altogether whenever the chemical reaction cannot take place, be it by specific inhibition or by the absence of an essential enzyme, in contrast to some solute-translocating systems, which after inhibition are still capable of facilitated diffusion. Since in the vectorial transport system, as presented above, the reaction of the solute with the carrier and its translocation cannot be clearly separated from each other, the distinction be-tween affinity effects and velocity effects of the coupling loses its meaning here.

The phosphotransferase system is established for some, but not all, microbial transport systems. There are still transport pro-cesses in which no intermediate covalent bond can be found and which can proceed without the operation of the phosphate-trans-ferring system described above. So in some yeasts inducible active transport systems of certain sugars have been described, which in the absence of this induction function as a facilitated, i.e., carrier-mediated transport system. Also the transport of lactose and other β-galactosides in *E. coli* can proceed without a phos-phate-transferring system. So it appears that many microbial systems of active transport can hardly be explained otherwise than by a "classical" carrier transport mechanism. This especial-ly holds for mammalian transport systems. It has been clearly demonstrated for sugar transport in the intestine that interme-diate phosphorylation of the transported sugar does not occur, since sugars may still be transported after all the OH groups found to be susceptible to phosphorylation have been blocked.

6.4.6 Glutamyl-Cycle. The "glutamyl" cycle, proposed by Meister and Orlowsky (MEISTER, 1973) to account for active amino acid trans-port in some animal tissues, would belong into the same category with the P-transferase system. Here the amino acid to be trans-ported is assumed to be glutamylated by a reaction with glutathion "across" the membrane, so that the amino acid appears first in the glutamylated form inside the cell. In a later step, the free amino acid is split off and the glutamyl residue utilized for the stepwise synthesis of glutathion. The analogy to the PTS system is obvious: The glutamyl amino acid corresponds to the sugar phosphate, and the glutathion, to the phosphoryl donator. It is easy to devise vectorial enzyme systems similar to those depicted in Figures 22 and 23 with corresponding channels spe-cific for the free amino acid and the glutamyl product of the same amino acid.

Whether and to what extent amino acids are transported by the glutamyl system is still controversial. It is generally believed that the glutamate cycle has no place wherever amino acids are transported by secondary active systems, by Na^+ or H^+ gradients. Furthermore, it has meanwhile been reported that in fibroblasts of an individual suffering from an inborn lack of glutamyl trans-ferase, the systems of amino acid transport was unimpaired (SCHULMAN et al., 1975; PELLEFIGUE et al., 1976).

6.5 Treatment of Solute-Specific Coupling in Terms of the Law of Mass Action (LMA)

6.5.1 General Procedure. The treatment of primary and secondary active transport in terms of the LMA will be essentially an extension of that of facilitated diffusion, and will accordingly be based on the same simplified model, which, as we recall, was assumed to consist of the following four distinct steps arranged cyclically in series (see Fig. 14):

Step 1. The binding of the transported solute to the carrier at the cis side of the barrier.

Step 2. The translocation of the carrier solute complex through the barrier. Also here we need not make any commitment as to whether this translocation is brought about by a translatory or rotary movement, or simply by a change in conformation of the carrier molecule.

Step 3. The release of the solute from the carrier site at the trans side of the barrier.

Step 4. The restoration step, which brings the empty carrier back into the position and condition from which it started.

As mentioned before, in active transport at least one of these four steps has to be coupled to an exergonic process. This may be also a translocation process, as in secondary active transport, or a chemical reaction, as in primary active transport. The equations that we shall derive for this model are analogous to those used in facilitated diffusion, but have to include the processes involved in coupling.

Since on the adopted level of simplification the fundamental differences between osmo-osmotic and chemi-osmotic coupling are small, it will be easy to derive from some general equations both those for osmo-osmotic coupling and those for chemi-osmotic coupling, a procedure that needs little more than minor modifications and changes of the symbols.

Coupling is in reality never complete, as some of the energy made available is always lost via leakage pathways. Leakage is primarily an energetic concept and will therefore be dealt with again later in connection with the thermodynamic treatment of coupling. To the extent, however, that leakages depend on mechanistic details of a transport system, they are also reflected in the kinetic transport equations based on the LMA. It will be instructive to include special leakage pathways in the underlying models, to trace them in the parameters of the transport equations, and to assess their presumable effects on the overall transport process. Before doing so we shall briefly describe and characterize the various kinds of leakages that are to be expected in every system of active transport.

6.5.2 Leakages. *Outer leakage* is caused by flows of the transported solute along pathways parallel to the pumping pathway. It allows dissipation, or at least diversion of energy stored in the electrochemical potential gradient of the solute transported. It is usually attributed to free, or at least uncoupled diffusion of the transported solute through pores or other leaky parts of the osmotic barrier. The decisive factor is that it is not coupled to the transport process under consideration, whereas it is unimportant whether it is instead coupled energetically to another process, so that some of the energy liberated by the leakage could do useful work. Such a flow would still have to be treated as leakage with respect to the process under consideration. For example, there may be several processes, osmotic and chemical ones, which are energized by the hydrolysis of ATP drawn from the same ATP pool. For each of these processes, all the other ones have to be considered as outer leakage.

Inner leakage, in contrast to outer leakage, concerns the carrier mechanism itself: It is due to imperfections that allow idling movements of the carrier. There are two major pathways for such leaks: (1) by the forward movement of the active, but unloaded carrier (slipping), and (2) by the backward movement of the inactive carrier of transported solute in combination with the inactivated carrier (backcycling). The first one becomes significant

Secondary active transport

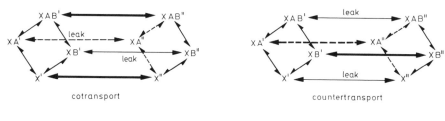

Primary active transport

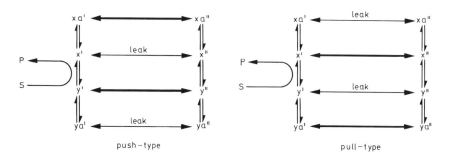

Fig. 25. Inversion of transport direction in active transport. *Fat arrows* represent transport effective steps, *thin ones* leakage steps. It can be seen that the same coupling can produce a push or a pull effect, depending on the relative rate of the steps involved

if cis concentration and mobility of the active unloaded carrier are high. The second one is the greater, the less complete the discharge of the transported solute on the trans side of the carrier, if for instance the carrier in its inactive form still has some residual affinity for the transported solute. The leakage pathways and their relation to the coupled pathways are depicted in Figure 25. To the extent that these leakage pathways are considered in the rate equations derived for any model of active transport, it will enable us to anticipate their effects on the parameters of these equations.

As a rule, we omit outer leakage in the first derivations in order to obtain a more lucid picture of the coupling in the transport mechanism as such.

Whereas the general procedure in developing rate equations for coupled transport systems, as has been pointed out, is analogous to that previously used for facilitated diffusion, we expect the equations of coupled transport to give one more important piece of information, namely, to tell us whether and to what extent the coupling leads to active transport. As a useful criterion of active transport we suggest the Haldane ratio (HR) (HALDANE, 1930). The HR can be derived from the rate equations by the subsequently derived procedures, and may also be tested experimentally under suitable conditions. It should differ from unity in case of active transport. It is also capable of indicating the significance of the various kinetic parameters, including those of leakage, for the effectiveness of the coupling device.

The HR can be obtained

1. By dividing the ratio $(^{o}J_{max}/K_m)$ for the initial net flow in the forward direction (a" = 0) by the corresponding ratio for the net flow in the backward direction (a' = 0) for the same concentration of b

$$ HR = \frac{\overset{\rightarrow}{^{o}J}_{max} \cdot K_m''}{K_m' \cdot \overset{\leftarrow}{^{o}J}_{max}} \tag{6.10a} $$

2. By determining the maximum flux ratio at level flow at vanishing activity of A (a' = a" → 0)

$$ HR = f_a^{max} \tag{6.10b} $$

3. By determining the maximum static head ratio:

$$ HR = \left(\frac{a''}{a'}\right)_{J_a=0}; \tag{6.10c} $$

Procedure 3 is the simplest one in the present context and will usually be applied in the following.

Since the HR is equal to the equilibrium constant,

$$ RT \ln \frac{\overset{\rightarrow}{^{o}J}_{max}}{\overset{\leftarrow}{^{o}J}_{max}} \frac{K_m''}{K_m'} = RT \ln K_{eq} = A_r^{o} \tag{6.11} $$

A_r^o being the standard driving force of the reaction concerned; RT ln (HR) should represent the effective nonconjugate driving effort in active transport. Hence the HR can be considered a link between the kinetics and the energetics of coupled transport.

6.5.3 Secondary Active Transport. We shall apply this procedure first to a model of secondary active transport, A being the transported solute and B, the cosolute, assuming that the transport is brought about by a ternary complex formed by the carrier and the two solutes A and B. This amounts to coupling between two flows, of which that of A is the "driven" one and that of B, the "driving" one. The distinction between a driven and a driving flow is arbitrary and implies here that A is to be transported uphill by the downhill flow of B. It could, of course, be the other way around, depending on the differences in electrochemical potential across the barrier for each solute.

As in facilitated diffusion, we start with the basic equations to obtain the general rate equation.

The conservation equation: Of the various carrier species we have now four on each side, the empty one (X), the two binary complexes (XA, XB), and the ternary complex (XAB). The equation is as follows:

$$x' + ax' + bx' + abx' + x'' + ax'' + bx'' + abx'' = x_T \qquad (6.12)$$

The steady-state equation: Giving each carrier species a special permeability constant (or probability of transition) we obtain

$$P_x x' + P_{ax} ax' + P_{bx} bx' + P_{abx} abx' = P_x x'' + P_{ax} ax'' + P_{bx} bx'' + P_{abx} abx''$$
$$(6.13)$$

The transport equation:

$$J_a = P_{ax} ax' + P_{abx} abx' - (P_{ax} ax'' + P_{abx} abx'') \qquad (6.14)$$

To solve these equations we make the following simplifying assumptions, which may not all be justified for real systems, but which improve the lucidity of the relationships without violating basic principles:

1. The movement of A and B in the free form is negligible, in other words, outer leakage can be disregarded.

2. The interactions between carrier in the interface and solute in adjacent solutions are very fast as compared to the translocation of the various carrier species, so that we can assume quasi-equilibrium between carrier species and solutes at the interface.

3. The system is symmetric. Hence the permeability and dissociation constants of each carrier species should be the same in either direction and on either side, respectively.

4. Both solutes A and B are electrically neutral. Hence we do not include any electric effects even though this is not realistic since the cosolute is most likely an ion (Na^+ or H^+) in secondary transport. In ion-linked cotransport, electric effects

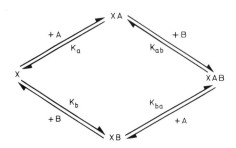

Fig. 26. Formation of ternary complex in secondary active transport. For details see text

will even introduce additional complications into the system, which have to be taken into account, as will be discussed later. In the present context, however, they are unnecessary to an understanding of the principle of coupling in cotransport and will therefore be disregarded here.

Having adopted the quasi-equilibrium treatment, we can replace each complex with the product of the two ligand concentrations, divided by the appropriate dissociation constant according to the following equations:

$$ax' = \frac{a' \cdot x'}{K_a}; \quad ax'' = \frac{a'' \cdot x''}{K_a}; \quad bx' = \frac{b' \cdot x'}{K_b}; \quad bx'' = \frac{b'' \cdot x''}{K_b} \tag{6.15}$$

For the ternary complex the following relationship should hold for the dissociation constants (Fig. 26):

$$\frac{K_a}{K_{ba}} = \frac{K_b}{K_{ab}} = r \tag{6.16}$$

In other words, the effects of the binding of one solute to the carrier on its affinity for the other are mutual. Accordingly, we obtain for the ternary complex

$$abx' = r \frac{a' \cdot b' \cdot x'}{K_a K_b} \qquad abx'' = r \frac{a'' \cdot b'' \cdot x''}{K_a K_b} \tag{6.17}$$

where r represents the "affinity factor" accounting for a change in affinity between binding site and ligand in a ternary complex as compared to the two binary complexes. To keep the number of different symbols small, we use normalized activities of A and B, i.e., $\alpha = (a/K_a)$ and $\beta = (b/K_b)$.

Inserting these values into the transport equation we obtain the final equation, which gives us the coupled flow of A as a function of the concentration of A or B on both sides of the membrane:

$$J_a = \frac{x_T}{\Sigma} \left[(P_{ax} + r\,P_{abx}\beta') \cdot (P_x + P_{bx}\beta'') \, \alpha' \right. \tag{6.18}$$

$$\left. - (P_{ax} + r\,P_{abx}\beta'') \cdot (P_x + P_{bx}\beta') \, \alpha'' \right]$$

$$\Sigma = (1 + \alpha' + \beta' + r\alpha'\beta')(P_x + P_{ax}\alpha'' + P_{bx}\beta'' + r\,P_{abx}\alpha''\beta'')$$

$$+ (1 + \alpha'' + \beta'' + r\alpha''\beta'')(P_x + P_{ax}\alpha' + P_{bx}\beta' + r\,P_{abx}\alpha'\beta')$$

The equation combines both cotransport and countertransport, as can be seen after rearranging the numerator and setting $\alpha' = \alpha''$ (level flow):

$$J_a = \frac{x_{T\alpha}}{\Sigma}\left[r\cdot P_x \cdot P_{abx}(\beta'-\beta'') + P_{ax}\cdot P_{bx}(\beta''-\beta')\right] \tag{6.19}$$

Of the terms within the brackets, $r\,P_x\cdot P_{abx}(\beta'-\beta'')$ represents cotransport, and $P_{ax}\cdot P_{bx}(\beta''-\beta')$ represents countertransport. The direction of the overall transport depends on which of the products $r\cdot P_x\cdot P_{abx}$ or $P_{ax}\cdot P_{bx}$ is the greater one. If $r\cdot P_x\cdot P_{abx} > P_{ax}\cdot P_{bx}$ we have predominantly cotransport; whereas in the opposite case, if $r\cdot P_x\cdot P_{abx} < P_{ax}\cdot P_{bx}$ we have predominantly countertransport. If the products are equal, there is no active transport at all. Obviously this model can serve for both cotransport and countertransport, depending on the relative rates and binding affinities of the various carrier species. Each of the terms is an antagonist of the other. In other words: In a cotransport system, the countertransport term is a leakage whereas a countertransport system, the cotransport term is a leakage.

The *unidirectional fluxes* of this transport are greater than the corresponding net fluxes:

$$\overrightarrow{J_a} = \frac{x_T}{\Sigma}(P_{ax} + r\cdot P_{abx}\beta')\cdot(P_x + P_{ax}\alpha'' + P_{bx}\beta'' + r\cdot P_{abx}\alpha''\beta'')\,\alpha'$$

$$\tag{6.20a}$$

and

$$\overleftarrow{J_a} = \frac{x_T}{\Sigma}(P_{ax} + r\cdot P_{abx}\beta'')\cdot(P_x + P_{ax}\alpha'\ P_{bx}\beta' + r\cdot P_{abx}\alpha'\beta')\,\alpha''$$

$$\tag{6.20b}$$

Hence we expect negative tracer coupling (isotope interaction) for this system.

To test whether this system is active we also derive the HR: To get a more general picture we do not make special assumptions as to the concentrations of A and B here. The simplest way to obtain the HR is by the third of the suggested methods, namely, from maximum static head ratio, i.e., the overall net rate equation by setting $J_a = 0$. We obtain

$$\left(\frac{\alpha''}{\alpha'}\right)_{J_a=0} = HR = \frac{(P_{ax} + r\cdot P_{abx}\cdot\beta')(P_x + P_{bx}\beta'')}{(P_{ax} + r\cdot P_{abx}\cdot\beta'')(P_x + P_{bx}\beta')} \tag{6.21a}$$

To investigate the relation between HR and the various parameters, we solve the parentheses:

$$HR = \frac{P_x P_{ax} + P_{ax}P_{bx}\beta'' + r\,P_x P_{abx}\beta' + r\,P_{bx}P_{abx}\beta'\beta''}{P_x P_{ax} + P_{ax}P_{bx}\beta' + r\,P_x P_{abx}\beta'' + r\,P_{bx}P_{abx}\beta'\beta''} \tag{6.21b}$$

It is seen that HR reflects the two opposing tendencies, represented by terms for cotransport and countertransport, respectively. To obtain pure cotransport we assume that

$$r \cdot P_x \cdot P_{abx} \gg P_{ax} \cdot P_{bx}$$

so that the terms containing $P_{ax} \cdot P_{bx}$ can be neglected. The ratio reduces to

$$(HR)_O = \frac{P_x P_{ax} + r\, P_{abx}(P_{bx}\beta'' + P_x)\beta'}{P_x P_{ax} + r\, P_{abx}(P_{bx}\beta' + P_x)\beta''} \quad \text{(cotransport)} \quad (6.21c)$$

Even though outer leakage has been disregarded, (HR) does not reach the theoretical maximum (β'/β''), as it would in true equilibrium, because of inner leakage. In the HR we recognize the terms that account for inner leakage and thus prevent HR from becoming equal to (β'/β'') : $P_x P_{ax}$ and the terms with $r \cdot P_{abx} \cdot P_{bx}$. The first represents leakage by backcycling of A via XA, which is determined mainly by P_{ax}, and the second, leakage by slipping, i.e., by movement of transport-active carrier (XB) without transporting A, which is determined mainly by P_{bx}. Only if these leakage terms are abolished, i.e. by making P_{ax} and $P_{bx} \to O$, does HR approach the theoretical maximum value $(\beta'/\beta'')=(b'/b'')$.

The same model can be converted into one of pure countertransport simply by assuming that $P_{ax} \cdot P_{bx} \gg P_x \cdot P_{abx}$ so that terms containing the latter product can be neglected. We then get

$$HR = \frac{P_x P_{ax} + (r\, P_{bx} P_{abx}\beta' + P_{ax}P_{bx})\beta''}{P_x P_{ax} + (r\, P_{bx} P_{abx}\beta'' + P_{ax}P_{bx})\beta'} \quad (6.22)$$

Also here the theoretical maximum value of HR, which is now (β''/β'), is not reached because of the terms of the inner leakage terms. One of them is the same as in cotransport, $P_x P_{ax}$, but in the present context it represents *slipping*, which is determined by P_x. The term $r\, P_{bx} P_{abx} \beta' \beta''$ now represents *backcycling*, and is determined by P_{abx}, i.e., by the mobility of the ternary complex (ABX).

The mechanistic aspects of inner leakage are visualized for each co- and countertransport in Figure 25, upper part. We see that in cotransport the slipping occurs along pathway BX' → BX" and the backcycling, along pathway AX" → AX', whereas the effectiveness of coupling depends on the rates of the pathways ABX' → ABX" and of X" → X', of the ternary complex, and of the unloaded inactivated carrier, respectively. With countertransport the situation is reversed: Slipping and backcycling occur along the pathways ABX' ← ABX" and X' → X', respectively, whereas the pathways AX' → AX" and BX" → BX' account for the effectiveness of coupling

It also follows from the above equations that HR remains finite for either transport direction even if the trans concentration of B becomes zero so that β'/β" or β''/β', respectively, and hence the ultimate driving force, become infinite. The reason for this seeming contradiction between the kinetic and energetic treatment is connected in either system with the inner leakage due to slipping for the cotransport ($P_{ax} = 0$) or backcycling for the countertransport ($P_x = 0$). By abolishing the corresponding leakage pathways the HR reduces to

$$(HR)'_{P_{ax} \to 0} = \frac{(P_x + P_{bx} \beta'') \beta'}{(P_x + P_{bx} \beta') \beta''} \quad \text{for cotransport} \quad (6.23a)$$

and

$$(HR)''_{P_x \to 0} = \frac{(r \, P_{abx} \beta' + P_{ax}) \beta''}{(r \, P_{abx} \beta'' + P_{ax}) \beta'} \quad \text{for countertransport} \quad (6.23b)$$

In either case HR also becomes infinite if the driving force is infinite.

Obviously the inner leakage via AX reduces the effective driving force to such an extent as to bring it into a finite range, even if the intrinsic driving force is infinite. The equation also shows that HR approaches unity, i.e., that no active transport is possible if the unloaded carrier is immobile, i.e., if $P_x \to 0$ in cotransport, or $P_{bx} \to 0$ in countertransport.

Even though none of the two pathways of inner leakage need be negligible, the product of the rate coefficients that account for these leakages must be very small. Otherwise the inversion of the transport becomes appreciable, which would render the transport system extremely ineffective and wasteful. Hence it is reasonable to presume that if one of the two leakage parameters is appreciable, the other must be negligible. Hence in cotransport P_{ax} and P_{bx}, in countertransport P_x and P_{abx}, must be very small.

Outer leakage. It is noteworthy that the HR expressions above do not contain the concentration of A. In other words, the depressing effect of inner leakage on HR as compared to (β''/β') is independent of a' and a". This is quite different with outer leakage, which so far has been disregarded in our models. To include outer leakage, the net transport has to be expanded by the term P_a (a' - a") = $P_a K_a$ ($\alpha' - \alpha''$), in which P_a stands for the uncoupled permeability coefficient of A. The equations become much more cumbersome, because the HR gets an additional term in both numerator and denominator, namely,

$$\frac{P_a \, K_a \, \Sigma}{x_T} = \Lambda_0 (a)$$

Σ being the same as in the general rate equation. This new term still further depresses HR as compared to the equilibrium value. Since Σ contains terms with α' and α'', the HR is no longer independent of the concentrations of A: It becomes smaller with increasing concentrations of A, as appears to be in agreement with experimental observations. At vanishing concentrations of A, however, the term of outer leakage does not disappear, because Σ does not become zero under these conditions. $\left(\frac{a''}{a'}\right)_{J_a=0}$, the

static head ratio, depends on outer leakage in the same way as does HR, i.e., it depends on the concentration of A only in the presence of outer leakage. In contrast, the flux ratio of level flow also depends on the concentration of A in the absence of outer leakage, so that it becomes identical with HR and $\left(\dfrac{a''}{a'}\right)_{J_a=0}$ at vanishing A under all conditions.

In summary the HR can be given the general form

$$HR = \frac{M' + \Lambda_i^s + \Lambda_i^b + \Lambda_o(a)}{M'' + \Lambda_i^s + \Lambda_i^b + \Lambda_o(a)} \tag{6.24}$$

in which M' and M" stand for $r \cdot P_{abx} \cdot P_x \cdot \beta'$ and $r \cdot P_{abx} \cdot P_x \cdot \beta''$, respectively, in cotransport; and for $P_{ax} P_{bx} \beta''$ and $P_{ax} P_{bx} \beta'$, respectively, in countertransport. Λ_i^s and Λ_i^b stand for the terms of inner leakage by slipping and backcycling, respectively, whereas $\Lambda_o(a)$ stands for the term of outer leakage, which can be a function of a' and a", the concentration of A, the transported solute.

As stated previously, transport systems are usually described in terms of Michaelis-Menten kinetics, on the assumption that the initial net flow, i.e., the flow at zero trans concentration of the transported solute (A), can be described by the equation

$$J_a = \overset{o}{J}_{max} \frac{a'}{K_m + a'}$$

This assumption is often confirmed experimentally with sufficient approximation. It can be shown that the rate equation derived above for secondary active transport can be reduced to a Michaelis Menten equation if the trans concentrations of A and B are set to zero. Under these conditions the initial net flow can be described in terms of the "standard parameters", i.e., $\overset{o}{J}^a_{max}$, the maximum rate, and K_m^a (Michaelis constant) of the initial net forward flow of A. It is interesting to see how these parameters relate to the kinetic coefficients and to the concentration of the cosubstrate B.

$$\overset{o}{J}^{max}_a = x_T \frac{P_x (P_{ax} + r P_{abx} \beta')}{P_x + P_{ax} + (P_x + P_{abx}) r \beta'} \quad (\text{at } a'' = o, \ b'' = o) \tag{6.25}$$

$$K_m = K_a \frac{2P_x + (P_x + P_{bx}) \beta'}{P_x + P_{ax} + (P_x + P_{abx}) r \beta'} \quad (\text{at } a'' = o, \ b'' = o) \tag{6.26}$$

It is easy to reduce these parameters to those of a pure affinity-type model or a pure velocity-type model, respectively.

For the affinity-type model we let all P_i be equal to P_x. Consequently,

$$\overset{o}{J}^{max}_a = \frac{P_x \cdot x_T}{2} \tag{6.27}$$

and

$$K_m = K_a \frac{1 + \beta'}{1 + r\beta'} \tag{6.28}$$

The corresponding parameters for the velocity type, in which r is set equal unity, are

$$\mathcal{J}_a^{max} = x_T \cdot \frac{P_x (P_{ax} + P_{abx} \beta')}{P_x + P_{ax} + (P_x + P_{abx}) \beta'} \qquad (6.29)$$

and

$$K_m = K_a \frac{2 P_x + (P_x + P_{bx}) \beta'}{P_x + P_{ax} + (P_x + P_{abx}) \beta'} \qquad (6.30)$$

It can be seen that the standard parameters of this equation, the maximum flow ($\mathcal{J}_a^{max}$) and the half saturation constant (K_m), are somewhat involved functions of the kinetic parameters and of the concentration of the driver solute as well.

If the above derivation of the kinetic equations for co- and countertransport of the various types, no special consideration is given to the influence of an *electrical potential* across the cellular membrane. Since the driver solute is as a rule an electrolyte ion (Na^+, H^+), co- and countertransport must obviously also depend on the electrical potential, regardless of whether the ternary complex has an electrical charge or not. If the empty carrier has no charge, then the ternary complex must have one equivalent to the number of Na^+ or H^+ bound. On the other hand, the ternary complex can be neutral only if the empty carrier must have a negative charge for each cation it can bind. Hence, at least one of the two opposing translocation steps is associated with a charge transfer and must therefore depend on the electrical field.

In the equations of irreversible thermodynamics the electrical PD is usually not explicitly noted because it is already implicity incorporated in the electrochemical PD of the driving ion. This is certainly correct from the energetic point of view, since an electrical PD and a chemical PD should be equivalent and interchangeable as a driving force. It is, however, not necessarily correct kinetically, i.e., if transport rates are concerned. These rates may not remain the same if for instance an electrical PD is applied to replace an equivalent difference in chemical potential.

Not each of the steps shown in our diagram (Fig. 14) need be affected by the electric PD. For instance, the binding and release of the ligands by the carrier should not be, whereas the translocation of the charged carrier species is electrically sensitive (rheogenic). Whether it is the forward movement of the ternary complex or the backward movement of the empty carrier that is directly driven by an electric potential depends on the charge of the empty carrier itself, as discussed above. The influence of the electric potential field on the overall transport depends therefore on the extent to which the electrically sensitive (rheogenic) step is rate limiting. As a rule, we do not know which is the rate-limiting step, but from the effect of electric potential on the overall process some information might be gained in this respect. So far, however, little has been done in this direction.

Source and sink system

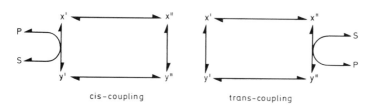

cis-coupling   trans-coupling

<u>Fig. 27.</u> Source-and-sink system. Schematic representation of cis coupling and trans coupling, respectively

A more detailed theoretical investigation about the effect of the electric potential was recently carried out (GECK and HEINZ, 1976; HEINZ and GECK, 1977). In addition, the electric PD may have some unpredictable effects on the permeability of the membrane concerned, effects which cannot a priori be considered in this theoretical treatment.

<u>6.5.4 *Primary Active Solute Translocation.*</u> To develop the analogous equation for the chemi-osmotic coupling, e.g., for the source-and-sink system, we distinguish between two forms of the carrier, X and Y. Unless stated otherwise, we shall define X as the "active" form, which is to have a higher affinity for the substrate than has Y, the "inactive" form (Fig. 27).

In order for active transport to occur, either the conversion of X to Y or that of Y to X has to be coupled to a chemical reaction, let us say to the transformation of a substrate S into the product P:

If coupling occurs at step 1 on the cis side (') according to the reaction

$$Y' + S \rightleftharpoons X' + P \tag{6.31}$$

the transition of X to Y (step 3) on the trans side (") must occur without coupling

$$X'' \rightleftharpoons Y''$$

If coupling occurs at the trans side the converse is true: The formation of X from Y is not coupled whereas the inactivation of X on the trans side is coupled S → P

$$X'' + S \rightleftharpoons Y'' + P \tag{6.32}$$

The two reactions, activation and inactivation of the carrier, no matter which of them is coupled, must be strictly separated spatially, as might be achieved by appropriate location of the two enzymes required for the coupled, and for the uncoupled reaction, respectively, at the different sides of the barrier.

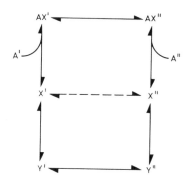

Fig. 28. Carrier pathways in primary active transport. For details see text

The complete cycle may be depicted as is shown in Figure 28.

To derive the rate equations we make the same simplifying assumptions as previously, namely, that outer leakage is absent, that the rate is limited by the translocation steps (quasi-equilibrium) and that the system is symmetric. The quasi-equilibrium is extended here also to both the coupled and uncoupled transformation between X and Y, so that only the translocation steps are rate limiting.

The assumptions of quasi-equilibrium and symmetry permit the following equilibrium relationships, which apply to both cis and trans coupling

$$\frac{a' \cdot x'}{ax'} = \frac{a'' \cdot x''}{ax''} = K_{ax} \qquad (6.33a)$$

$$\frac{a' \cdot y'}{ay'} = \frac{a'' \cdot y''}{ay''} = K_{ay} \qquad (6.33b)$$

The transition X and Y, as mentioned, occurs by different pathways on the two sides, one being coupled to the driver process (S → P), the other not. Since for either transition we are assuming quasi-equilibrium, we can postulate the following relationships:

1. For the cis coupling

$$\frac{x' \cdot p}{y' \cdot s} = \frac{K_{sp}}{K_{xy}} \quad \text{and} \quad \frac{y''}{x''} = K_{xy} \qquad (6.34)$$

K_{sp} and K_{xy} being the equilibrium constants of the driver reaction (S → P) and of the interconversion between the two carrier forms. Accordingly, for cis coupling we can approximately set

$$y' = \frac{p}{s} \cdot \frac{K_{xy}}{K_{sp}} x', \quad \text{or} \quad = \frac{K_{xy}}{\Gamma} x' \quad \text{and} \quad y'' = K_{xy}x'' \qquad (6.35)$$

Γ indicates how far away the driving chemical reaction is from its equilibrium. It is a function of the affinity available from this reaction (S → P) and may be called "reactivity coefficient".

$$\Gamma = e^{\frac{A_{ch}}{RT}} = \frac{s}{p} K_{sp} \tag{6.36}$$

The direction of active solute transport with any given model depends on whether Γ is greater or smaller than unity. Unless stated otherwise, Γ is always considered either unity or greater than unity in our equations.

2. For the trans coupling

$$\frac{y'}{x'} = K_{xy} \text{ (left side)} \quad \text{and} \quad \frac{y'' \cdot p''}{x'' \cdot s''} = K_{xy} \cdot K_{sp} \tag{6.37}$$

so that

$$y' = K_{xy} \cdot x' \quad \text{and} \quad y'' = K_{sp} \cdot \frac{s''}{p''} \cdot K_{xy} \cdot x'' = \Gamma K_{xy} x'' \tag{6.38}$$

We now develop the basic equations as previously. The conservation equation,

$$x' + y' + ax' + ay' + x'' + y'' + ax'' + ay'' = x_T, \tag{6.39}$$

and the steady state equation,

$$P_x \cdot x' + P_y \cdot y' + P_{ax} \cdot ax' + P_{ay} \cdot ay' = P_x x'' + P_y y'' + P_{ax} ax'' + P_{ay} ay'' \tag{6.40}$$

and the overall transport equation,

$$J_a = P_{ax}(ax' - ax'') + P_{ay}(ay' - ay'') \tag{6.41}$$

Inserting the above equilibrium expressions we can derive from these equations the general rate equation, giving the rate of the transport of the solute concerned, J_a, as a function of the total number of carrier species (x_T), the activities of the transported and cotransported solutes, of the reactants and products of the driving reactions, and the appropriate equilibrium and rate (or probability) constants.

Thus we obtain for the *cis coupling*, using the normalized concentrations of solute A, $\alpha' = \dfrac{a'}{K_{ay}}$ and $\alpha'' = \dfrac{a''}{K_{ay}}$:

$$J_a = \frac{x_T}{\Sigma} \left[(P_x + P_y K_{xy})(r\Gamma P_{ax} + P_{ay} K_{xy}) \alpha' - (\Gamma P_x + P_y K_{xy})(r P_{ax} + P_{ay} K_{xy}) \alpha'' \right]$$

$$\Sigma = \left[(\Gamma + K_{xy}) + (r\Gamma + K_{xy}) \alpha' \right] \left[(P_x + P_y K_{xy}) + (r P_{ax} + P_{ay} K_{xy}) \alpha'' \right]$$

$$+ \left[(1 + K_{xy}) + (r + K_{xy}) \alpha'' \right] \left[(\Gamma P_x + P_y K_{xy}) + (r\Gamma P_{ax} + P_{ay} K_{xy}) \alpha' \right]$$

$$\Gamma = \frac{s'}{p'} \cdot K_{sp}, \quad r = \frac{K_{ay}}{K_{ax}} \tag{6.42}$$

In analogy to secondary active transport, the general rate equation for the cis coupling includes both a *push* mechanism (toward the right) and a *pull* mechanism (toward the left). Which of these effects predominates, and thus determines the overall transport direction, depends merely on the relative magnitude of certain rate and dissociation constants, as will become clearer from the HR farther below.

The standard parameters, extracted from these equations for the initial net flow (a" = o), are:

$$K_m = K_{ay} \frac{(\Gamma + K_{xy})(P_x + P_y K_{xy}) + (1 + K_{xy})(\Gamma P_x + P_y K_{xy})}{(r\Gamma + K_{xy})(P_x + P_y K_{xy}) + (1 + K_{xy})(r\Gamma P_{ax} + P_{ay} K_{xy})} \qquad (6.43)$$

$$\overset{max}{\underset{a}{\mathcal{J}}} = \frac{x_T(P_x + P_y K_{xy})(r\Gamma P_{ax} + P_{ay} K_{xy})}{(r\Gamma + K_{xy})(P_x + P_y K_{xy}) + (1 + K_{xy})(r\Gamma P_{ax} + P_{ay} K_{xy})} \qquad (6.44)$$

To test whether there is active transport we determine the HR by any of the previously indicated methods:

$$HR = \frac{(P_x + P_y K_{xy}) \cdot (r\Gamma P_{ax} + P_{ay} K_{xy})}{(\Gamma P_x + P_y K_{xy}) \cdot (r P_{ax} + P_{ay} K_{xy})} \qquad (6.45)$$

Whether the system pushes A to the right side or pulls A to the left side depends on the magnitude of $r \cdot P_{ax}$ relative to $P_{ay} K_{xy}$. If $r P_{ax} \gg P_{ay} K_{xy}$, it pulls A to the left side. The smaller of the two products can be considered a "leakage" for the particular system, reducing the effectiveness of the coupling by permitting "back cycling" of the transported solute A via the inactive carrier. This can be seen from the HR, if for the sake of simplicity the backcycling terms are neglected. Thus, if in the push type ($r P_{ax} \gg P_{ay} K_{sy}$), HR is greater than unity:

$$HR = \frac{(P_x + P_y K_{xy})\Gamma}{\Gamma P_x + P_y K_{xy}} \qquad (6.46)$$

Since P_x, the coefficient of *slipping* for empty activated carrier, represents an additional leakage pathway, its depression ($P_x \to o$) will raise the accumulation ratio toward Γ, the theoretical equilibrium value.

If, on the other hand, $r \cdot P_{ax} \ll P_{ay} \cdot K_{xy}$, we obtain the pull type as indicated by a HR smaller than 1:

$$HR = \frac{P_x + P_y K_{xy}}{\Gamma P_x + P_y K_{xy}} \qquad (6.47)$$

i.e., the direction of active transport has been inverted. Since Y is now the active carrier, the leakage by slipping is determined by $P_y \cdot K_{xy}$. If it is very small compared to P_{xy}, the (inverse) accumulation ratio approaches $(1/\Gamma)$, the theoretical equilibrium value under these conditions.

In the pure *affinity type*, all Ps are taken to be equal, so that, in the absence of external leakage, the HR, or the static head accumulation ratio, is for the push type

$$HR = \frac{(1 + K_{xy})(r\Gamma + K_{xy})}{(\Gamma + K_{xy})(r + K_{xy})} \tag{6.48}$$

In the pure *velocity type*, r is unity so that

$$HR = \frac{(P_x + P_y K_{xy})(\Gamma P_{ax} + P_{ay} K_{xy})}{(\Gamma P_x + P_y K_{xy})(P_x + P_{ay} K_{xy})} \tag{6.49}$$

In analogy to secondary active transport, accumulation of A will occur only if

$$\frac{P_{ax} \cdot P_y}{P_{ay} \cdot P_x} > 1;$$

For the *trans coupling*, analogous equations can be derived with the difference that on the cis side $y' = K_{xy} \cdot x'$ and that on the trans side, owing to the coupling, $y'' = \Gamma K_{xy} x''$. We now get for the net flux

$$J_a = \frac{x_T}{\Sigma} \left[(P_x + \Gamma P_y K_{xy})(r P_{ax} + P_{ay} K_{xy}) \alpha' - (P_x + P_y K_{xy})(r P_{ax} + \Gamma P_{ay} K_{xy}) \alpha'' \right]$$

$$\Sigma = \left[(1 + K_{xy}) + (r + K_{xy}) \alpha' \right] \left[(P_x + \Gamma P_y K_{xy}) + (r P_{ax} + \Gamma P_{ay} K_{xy}) \alpha'' \right]$$

$$+ \left[(1 + \Gamma K_{xy}) + (r + \Gamma \cdot K_{xy}) \alpha'' \right] \left[(P_x + P_y K_{xy}) + (r P_{ax} + P_{ay} K_{xy}) \alpha' \right] \tag{6.50}$$

and for the standard parameters of initial net flow ($\alpha'' = o$)

$$K_m = K_{ay} \frac{(P_x + \Gamma P_y K_{xy})(1 + K_{xy}) + (P_x + P_y K_{xy})(1 + \Gamma K_{xy})}{(P_x + \Gamma P_y K_{xy})(r + K_{xy}) + (r P_{ax} + P_{ay} K_{xy})(1 + \Gamma K_{xy})} \tag{6.51}$$

$$J_{max} = x_T \frac{(r P_{ax} + P_{ay} K_{xy})(P_x + \Gamma P_y K_{xy})}{(P_x + \Gamma P_y K_{xy})(r + K_{xy}) + (r P_{ax} + P_{ay} K_{xy})(1 + \Gamma K_{xy})} \tag{6.52}$$

The HR is

$$HR = \frac{(P_x + \Gamma P_y K_{xy})(r P_{ax} + P_{ay} K_{xy})}{(P_x + P_y K_{xy})(r P_{ax} + \Gamma P_{ay} K_{xy})} \tag{6.53}$$

If $r P_{ax} \gg P_{ay} K_{xy}$, we now have a *trans-coupled pull* mechanism driving A towards the right side, provided $\Gamma > 1$.

$$HR = \frac{P_x + \Gamma P_y K_{xy}}{P_x + P_y K_{xy}}$$

and if the slipping becomes negligible ($P_x \to o$)

$$HR \to \Gamma$$

as in the cis coupled push type.

On the other hand, if $rP_{ax} \ll P_{ay}K_{xy}$, be it because the affinity of A for Y exceeds that for X $(r < 1)$, or be it because AY is translocated much faster than AX $(K_{xy} \cdot P_{ay} \gg P_{ax})$, the direction of active transport is inverted:

$$HR = \frac{P_x + \Gamma P_y K_{xy}}{(P_x + P_y K_{xy})\Gamma} \tag{6.55}$$

We have now a *trans-coupled push* mechanism, driving A toward the left side at the same Γ. Since Y is now the active carrier form, its slipping, which depends on $P_y \cdot K_{xy}$, has to vanish in order to permit maximal transport efficiency in the inverse direction

$$(HR)_{P_y K_{xy} \to 0} \to \frac{1}{\Gamma}$$

We see that in analogy to secondary active transport, beside the neglected diffusional (outer) leakage of the solute A through the barrier, two different pathways of inner leakage, inherent in the carrier system, may be present here: The movement of unloaded active carrier (slipping), and the back flow of A in combination with the inactive carrier (back cycling). The term leakage, however, has proved to be relative, since which pathway is a leakage pathway and which is a coupling-effective pathway depend on whether the model functions by a pull or by a push mechanism. In the Haldane ratios, the two pathways are clearly identifiable.

As we have seen with secondary active transport, the Haldane ratios do not contain the concentrations of the transportee, hence the effect of inner leakage on its value does not depend on these concentrations. In other words, the maximum static head accumulation ratio should be the same for lower or higher concentrations of the solute.

This changes, of course, if outer leakage is included as well. As with secondary active transport, outer leakage would be included by expanding both numerator and denominator of the HR by the term:

$$P_a \cdot K_a \cdot \frac{\Sigma}{x_T}$$

P_a being the coefficient of uncoupled penetration of A. The factor Σ, the same as in Eq. (6.50), being a function of both solute concentrations, introduces a dependence of the static head accumulation ratio on these concentrations, as was the case with outer leakage in secondary active transport.

6.5.5 Combined Push and Pull Effects. The distinction we have made between push and pull effects is arbitrarily chosen with reference to the movement of the transported solute (A). With reference to

114

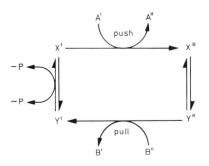

Fig. 29. The pull-push system of active antiport. A single coupling provides the energy for moving solute A from *left* to *right (push)* and of solute B from *right* to *left (pull)*. For details see text

the carrier movement, however, the same coupling on the same side can exert both a push and a pull effect at the same time, a push effect with respect to the carrier in the X form and a pull effect, with respect to the Y form of the carrier. If, for instance, the transition from Y to X (step 1) is energetically coupled whereas the reverse transition from X to Y on the other side is not (step 3), the same coupling reaction that "pushes" X from left to right must also have a "pull" effect on Y from right to left (Fig. 29).

This combined push and pull effect can be utilized simultaneously for the transport of two different solutes, let us say A and B, in opposite directions. We simply would have to postulate that the transition X → Y is associated with a change in affinity with respect to the two solutes: For instance, X, being transformed to Y, loses its affinity for A but instead gains in affinity for B, and vice versa. Such a system could well be imagined to be capable of actively transporting both A and B, but in opposite directions, thus functioning as an active antiporter system, such as the exchange pumps Na^+, K^+-ATPase, and possibly also the K^+-H^+-ATPase.

A simplified quantitative description of this model in terms of the LMA is possible by an extention of the analogous treatment of the uniport system, which could be either the push or the pull type. We need one additional assumption only, namely, that the interconversion between X and Y is associated with a drastic change in binding specificity. Accordingly we assume that X binds A more strongly than B, and that Y binds B more strongly than A. To keep the model simple and lucid we neglect any residual affinities, of X for B and Y for A, because these would merely reduce the efficiency of the system without changing basic principles. Accordingly, we expand the conservation and steady-state equations, respectively, by the terms for BY and the corresponding mobilities, omitting the terms containing AY.

1. $x' + y' + ax' + by' + x'' + y'' + ax'' + by'' = x_T$ (6.56)

2. $P_x \cdot x' + P_y \cdot y' + P_{ax} \cdot ax' + P_{bx} \cdot bx' = P_x \cdot x'' + P_y \cdot y'' + P_{ax} \cdot ax'' + P_{by} \cdot by''$

     (6.57)

3.   a) $J_a = P_{ax}(ax' - ax'')$;   b) $J_b = P_{bx}(bx' - bx'')$ (6.58)

As before, we assume that the transition $Y' \rightarrow X'$ is coupled to the chemical reaction $S \rightarrow P$, whereas the transition $X' \rightarrow Y''$ is not, and that both reactions are *quasi* at equilibrium. Hence we can write

$$y' = \frac{K_{xy}}{\Gamma} x' \quad \text{and} \quad y'' = K_{xy} x'' \tag{6.59}$$

Γ standing for $(s/p) \cdot K_{sp}$. We also assume equilibrium between the solute concentrations and their complexes with the carrier forms, and replace (a/K_{ax}) by α and (b/K_{by}) by β. Inserting these expressions in the above equations we finally obtain

$$J_a = P_{ax} x_T \cdot \frac{(P_x + P_y K_{xy} + P_{by} K_{xy} \beta'') \alpha' - (P_x + P_y \frac{K_{xy}}{\Gamma} + P_{by} \frac{K_{xy}}{\Gamma} \beta') \alpha''}{\Sigma} \tag{6.60a}$$

$$J_b = P_{by} x_T \cdot \frac{(P_x + P_y K_{xy} + P_{ax} \alpha'') \frac{K_{xy}}{\Gamma} \beta' - (P_x + P_y \frac{K_{xy}}{\Gamma} + P_{ax} \alpha') K_{xy} \beta''}{\Sigma} \tag{6.60b}$$

$$\Sigma = (1 + \frac{K_{xy}}{\Gamma} + \alpha' + \frac{K_{xy}}{\Gamma} \beta') (P_x + P_y K_{xy} + P_{ax} \alpha'' + P_{by} K_{xy} \beta'')$$

$$+ (1 + K_{xy} + \alpha'' + K_{xy} \beta'') (P_x + P_y \frac{K_{xy}}{\Gamma} + P_{ax} \alpha' + P_{by} \frac{K_{xy}}{\Gamma} \beta')$$

In the present context we are interested mainly in the extent to which the two transport processes are *active*: For this purpose we derive as before the flux ratios, each at level flow

$$f_a = \Gamma \frac{(P_x + P_{ax} \alpha) + K_{xy} (P_y + P_{by} \beta'')}{\Gamma (P_x + P_{ax} \alpha) + K_{xy} (P_y + P_{by} \beta')}, \quad \text{at } \alpha' = \alpha'', \ \beta' \neq \beta'' \tag{6.61a}$$

$$f_b = \frac{(P_x + P_{ax} \alpha'') + K_{xy} (P_y + P_{by} \beta)}{\Gamma (P_x + P_{ax} \alpha') + K_{xy} (P_y + P_{by} \beta)}, \quad \text{at } \beta' = \beta'', \ \alpha' \neq \alpha'' \tag{6.61b}$$

We see that the transport of each solute is stimulated by the opposing gradient of the other.

The activeness of both transports is tested quantitatively by the Haldane ratios

$$(HR)_a = \frac{\Gamma (P_x + K_{xy} P_y)}{\Gamma P_x + K_{xy} P_y} \tag{6.62a}$$

$$(HR)_b = \frac{P_x + K_{xy} P_y}{\Gamma P_x + K_{xy} P_y} \tag{6.62b}$$

We see that both transport processes are *active* according to definition, but in opposite directions. The ratio of the two

Haldane ratios at the same values of a', a" and b', b" in both equation is

$$\frac{(H\dot{R})_a}{(HR)_b} = \Gamma \tag{6.63}$$

As is to be expected, both transport processes have to share the same energy source. We also see that the transport of each solute is supported by an opposing gradient of the other solute.

6.5.6 Group Translocation. To get an idea of the quantitative relationships between a heterophasic (vectorial) reaction and group transfer in terms of the LMA we chose a system which comes close to that one might ascribe to the phosphotransferase system. We again make some simplifying assumptions, which may be unrealistic but which greatly facilitate understanding without sacrificing basic principles. We assume a single enzyme X with features outlined above, which after being activated by phosphorylation is capable of accepting a substrate A from the cis side ('). At the active center of the enzyme (or enzyme complex) the phosphoryl group is transferred to the substrate so that a new arrangement at the enzyme, the solute-phosphate complex being exposed to the trans side, emerges. The solute-phosphate moiety is then released and in a last step, split into the solute A" and free phosphate. This sequence of reaction steps can be visualized as follows:

1. $XP + A' \rightleftharpoons AXP'$
2. $AXP' \rightleftharpoons XAP''$
3. $XAP'' \rightleftharpoons X + AP''$
4. $X + C \sim P \rightleftharpoons XP + C$
5. $AP'' \rightleftharpoons A'' + P$

and arranged as a cycle (Fig. 30).

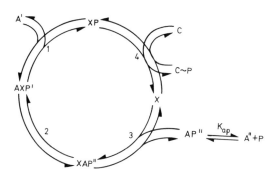

Fig. 30. The phosphotransferase system. Schematic diagram of the hypothetical reaction steps. For details see text

For the sake of simplicity, we assume that all steps that do not involve covalent bonds, i.e., in our scheme steps 1 and 3, are so fast that they can be treated as being in quasi-equilibrium. Hence step 2 and step 4, representing true chemical reactions, are slow enough to limit the overall reaction rate, with one exception: The hydrolysis of AP to A" + P is also assumed to be very fast compared to the other chemical reactions so that at any time free A" is close to equilibrium with its P-ester and free phosphate. Hence we can set

$$axp' = \frac{k_1}{k_{-1}}\, a' \cdot xp, \qquad xap'' = \frac{k_{-3}}{k_3} \cdot ap'' \cdot x;$$

or since

$$\frac{a'' \cdot p}{ap''} = K_{ap}, \qquad xap'' = \frac{k_{-3}}{k_3} \cdot \frac{a'' \cdot p \cdot x}{K_{ap}}$$

(6.64)

As previously, we assume that the system is in the steady state and that transport of A by a pathway other than via this system is negligible. Steady state is also assumed for $C \sim P$, C and P, which are considered to be maintained at constant values by metabolism.

In analogy to previous derivations we now write the basic equations (conservation, steady-state, and transport equation):

$$x + xp + axp' + xap'' = x_T$$

(6.65)

$$k_2 axp' + k_{-4} xp \cdot c = k_{-2} xap'' + k_4 x \cdot c \sim p;$$

(6.66)

$$J_a = k_2 axp' - k_{-2} xap''$$

(6.67)

After inserting the equilibrium expressions we can solve for x and xp, the latter taking here the function of the "active" form of the carrier, and then derive the overall rate equation for the net flux (J_a):

$$J_a = \frac{x_T}{\Sigma}\, (k_1 \cdot k_2 \cdot k_3 \cdot k_4\, K_{ap}\, c \sim p \cdot a' - k_{-1} k_{-2} k_{-3} k_{-4} \cdot c \cdot p \cdot a'')$$

(6.68)

According to the law of detailed balance $J_a = 0$ if a' = a" and $\frac{c \cdot p}{c \sim p} = K_{cp}$, the equilibrium constant of the reaction $C \sim P \rightleftharpoons C + P$ hence

$$k_1 k_2 k_3 k_4\, K_{ap} = k_{-1} k_{-2} k_{-3} k_{-4} \cdot K_{cp}$$

(6.69)

The net transport equation reduces to

$$J_a = \frac{x_T}{\Sigma}\, k_1 \cdot k_2 \cdot k_3 \cdot k_4\, K_{ap}\, (c \sim p \cdot a' - \frac{c \cdot p}{K_{cp}} \cdot a'')$$

(6.70)

$$= \frac{x_T}{\frac{K_{cp}}{c \cdot p}\,\Sigma}\, k_1 k_2 k_3 k_4\, K_{ap} \cdot (K_{cp} \cdot \frac{c \sim p}{c \cdot p}\, a' - a'').$$

If we set $K_{cp} \cdot \frac{c \sim p}{c \cdot p} = \Gamma$

$$J_a = \frac{x_T}{\frac{K_{cp}}{c \cdot p} \Sigma} \left[k_1 k_2 k_3 k_4 K_{ap} (\Gamma a' - a'') \right] \tag{6.71}$$

$$\frac{K_{cp}}{c \cdot p} \Sigma = k_{-1} k_3 K_{ap} (k_4 \Gamma + k_{-4} \frac{K_{cp}}{p}) + k_1 k_{-3} \frac{K_{cp}}{c} (k_{-2} + k_2) a' a''$$

$$+ k_1 k_3 K_{ap} (k_2 \frac{K_{cp}}{c \cdot p} + k_4 \Gamma) a' + k_{-1} k_{-3} \frac{K_{cp}}{c} (k_{-2} + k_{-4} c) a'$$

The standard parameters of net movement in the forward direction
(a" = 0) are

$$\overset{\rightarrow}{J_a^{max}} = x_T \frac{k_2 k_4 c \sim p}{k_2 K_{cp} + k_4 c \sim p} \qquad K'_m = \frac{k_{-1}}{k_1} \frac{k_4 c \sim p + k_{-4} c}{k_4 c \sim p + k_2} \tag{6.72a}$$

and in the backward direction (a' = 0)

$$\overset{\leftarrow}{J_a^{max}} = x_T \frac{k_{-2} k_{-4} \cdot c}{k_{-2} + k_{-4} \cdot c} \qquad K''_m = \frac{k_3}{k_{-3}} \frac{K_{ap} (k_{-4} c + k_4 c \sim p)}{p (k_{-4} c + k_{-2})} \tag{6.72b}$$

It is seen that $c \sim p$ is absolutely required for the initial net
flow in the forward direction, but not for that in the backward
direction, which instead depends on c and p. If
$\frac{c \cdot p}{c \sim p} > K_{cp}$ ($\Gamma < 1$) transport of A should occur in the reverse direction.

We also see that the system is intrinsically asymmetric, i.e.,
the asymmetry does not disappear if the driving force becomes
zero ($\frac{c \cdot p}{c \sim p} = K_{cp}$).

The coefficient of the asymmetry (at $\Gamma = 1$) is

$$Q = \frac{k_2 k_4 (k_{-2} + k_{-4} c) \cdot p}{k_{-2} k_{-4} (k_2 K_{cp} + k_4 \cdot c \cdot p)} \tag{6.73}$$

The unidirectional fluxes are greater than the initial net rates,
owing to a shuttling between the two forms of the active complex

$$axp' \longleftrightarrow xap''$$

Accordingly the flux ratio is

$$f_a = \frac{(k_1 k_2 k_3 k_4 K_{ap} \Gamma \cdot c \cdot p + k_1 k_2 k_{-2} k_{-3} K_{cp} \cdot p \cdot a'') a'}{(k_{-1} k_{-2} k_{-3} k_{-4} \cdot K_{cp} \cdot c \cdot p + k_1 k_2 k_{-2} k_{-3} \cdot K_{cp} \cdot p \cdot a') a''} \tag{6.74a}$$

$$= \frac{a'}{a''} \frac{(\Pi \cdot c \cdot p \Gamma + k_1 k_2 k_{-2} k_{-3} K_{cp} \cdot p \cdot a'')}{(\Pi \cdot c \cdot p + k_1 k_2 k_{-2} k_{-3} K_{cp} \cdot p \cdot a')} \tag{6.74b}$$

In agreement with what had been stated on asymmetry before, the
ratio of the unidirectional fluxes does not show the asymmetry.

Since $\left| \ln f_a \right| < \left| \ln \frac{a'}{a''} \right|$ there is negative tracer coupling. The Haldane ratio, derived from either the standard parameters or from the flux ratio at a', $a'' \to 0$ is

$$HR = \Gamma$$

as was to be expected, because no leakages are considered in the model.

7 Energetics of Coupled Transport
Treatment of Active Transport in Terms of Thermodynamics of Irreversible Processes (TIP)

7.1 The Basic Phenomenological Equations

7.1.1 General. To treat coupled transport in terms of TIP we can
proceed along two different paths: Firstly, we can start from
the TIP equations for uncoupled transport, as developed in the
previous section for free and facilitated diffusion, and expand
them to include the coupling effects. This is the conventional
and simpler way, which gives us the flows in terms of driving
forces and phenomenological coefficients only, but tells us littl
about the mechanistic relationships. Secondly, we can start from
the kinetic equations of the coupled processes, as derived in
the previous section for primary and ·secondary active transport
in terms of the LMA, and try to transform them into the corresponding thermodynamic equations. For reasons given in the previous section this transformation can be carried out only by the
use of certain approximations. The final equations arrived at
should be identical with those obtained by the first procedure,
but the second procedure enables us to interpret the phenomenological coefficients in terms of mechanistic and kinetic details of the underlying model. This has the advantage that the
range of linearity for the phenomenological coefficients can be
predicted, provided that the underlying mechanistic model is in
its essentials close enough to the real system.

In the following we shall outline both procedures, beginning
with the thermodynamic one (KATCHALSKY and CURRAN, 1965).

7.1.2 Conventional Approach. We recall that the uncoupled flow of a
given solute (i) can be equated to the conjugate driving force,
X_i, times the phenomenological coefficient L_{ii}

$$J_i = L_{ii} X_i;$$ (7.1)

or in the R notation,

$$X_i = R_{ii}^{\cdot} J_i.$$ (7.2)

It has been pointed out that in such a simple system in which
all forces other than the conjugate one can be neglected, $L_{ii}' = \frac{1}{R_{ii}}$, but that this relationship no longer holds in the presence
of nonconjugate driving forces.

Dealing now with several componénts we have first to link all
flows with their conjugate driving forces to obtain the rate of
the overall entropy production times the absolute temperature

$$T\dot{S} = J_i X_i + J_j X_j + \ldots J_k X_k + \ldots , \quad \dot{S} = \frac{dS}{dt} .$$ (7.3)

If Eq. (7.3) is complete and positive, we derive the individual flow equations as follows

$$J_i = L_{ii} \cdot X_i + L_{ij} \cdot X_j + L_{ik} \cdot X_k + \ldots \qquad (7.4a)$$

$$J_j = L_{ij} \cdot X_i + L_{jj} \cdot X_j + L_{jk} \cdot X_k + \ldots \qquad (7.4b)$$

$$J_k = L_{ki} \cdot X_i + L_{kj} \cdot X_j + L_{kk} \cdot X_k + \ldots \qquad (7.4c)$$

$\ldots\ldots$

Since according to Onsager $L_{ij} = L_{ji}$, $L_{ik} = L_{ki}$, $L_{jk} = L_{kj}$ etc.

$$J_i = L_{ii} X_i + L_{ij} X_j + L_{ik} X_k \ldots \qquad (7.5a)$$

$$J_j = L_{ij} X_i + L_{ij} X_j + L_{jk} X_k \ldots \qquad (7.5b)$$

$$J_k = L_{ik} X_i + L_{jk} X_j + L_{kk} X_k \ldots \qquad (7.5c)$$

X_j represents either the negative electrochemical potential (ECP) gradient of an additional species J or the affinity of a chemical reaction, to whose J_i is coupled, and L_{ij} is the corresponding (linear) cross coefficient. The conditions for the linearity of the coefficients are: (1) the overall system is close to equilibrium, and (2) the forces and affinities of all flows and reactions to which J_i is coupled are included.

The units of X are usually those of forces or affinities, respectively, but can also be different from these, depending on the unit of the corresponding flows. If, for instance, J is a volume flow, given in volume per time unit, X will have the dimension of a hydrostatic or osmotic pressure.

To better understand the advantages and limitations of this procedure, we shall illustrate the above relations by simple model systems in which only two flows, J_i and J_j, are coupled to each other. These flows could be those of two solutes of watery solutions, provided that solute-solvent interaction during membrane passage can be neglected.

As with uncoupled flows, two equivalent notations are commonly in use:

the L notation

$$J_i = L_{ii} X_i + L_{ij} X_j \qquad (7.6a)$$

$$J_j = L_{ji} X_i + L_{jj} X_j \qquad (7.6b)$$

the R notation

$$X_i = R_{ii} J_i + R_{ij} J_j \qquad (7.7a)$$

$$X_j = R_{ji} J_i + R_{jj} J_j \qquad (7.7b)$$

In contrast to uncoupled flows, L_{ii} is no longer equal to $(1/R_{ii})$.

For a simple system with only two coupled flows, the relation between L and the R coefficients is the following:

$$L_{ii} = \frac{R_{jj}}{R_{ii}\,R_{jj} - R_{ij}^{\,2}}; \quad L_{jj} = \frac{R_{ii}}{R_{ii}\,R_{jj} - R_{ij}^{\,2}}; \quad L_{ij} = -\frac{R_{ij}}{R_{ii}\,R_{jj} - R_{ij}^{\,2}}$$

(7.8)

The R notation lends itself to representation by an electric equivalent circuit, even if no electric charges are involved. It has the disadvantage that the cross coefficient, R_{ij}, is negative in positive coupling, e.g., in cotransport, and positive in negative coupling, e.g., in countertransport. With the L notation the signs of the cross coefficients, in contrast, correspond to the sign of the coupling. Both notations are general, applying to all kinds of interactions, that lead to coupling, provided that the above-mentioned conditions are met. They have proved most useful in treating coupling by frictional interaction, in particular in the joint movement of water and solutes across certain biological membranes.

7.1.3 Quasi-Chemical Notation. If the coupling involves a stoichiometric chemical reaction, such as the chemical combination between two transported solute species with one another, directly or by the mediation of a mobile carrier, the *quasi-chemical* notation may be more advantageous than the above one (HEINZ, 1974). This notation treats the overall transport process like a chemical reaction, dealing with reactants and products only. The reactants are represented by the solute species in the left compartment, and the products, by the same species in the right compartment. No attention is paid to what happens in between, i.e., during the passage across the membrane, except for the assumption that the coupled flows pass at a fixed stoichiometric ratio, for instance as a stoichiometric carrier-substrate complex. This notation seems especially useful if the mechanism of the transport process is not precisely known, but it loses its meaning if the interaction is nonstoichiometric, such as in coupling by friction.

The general procedure is in the following exemplified with two specific solutes, A and B. We start with the overall equation

$$\nu'_a \cdot A' + \nu'_b \; B' \;\Longleftrightarrow\; \nu''_a \; A'' + \nu''_b \; B''$$

(7.9)

The subscripts denote the individual species and the superscripts the compartment. The ν values give the stoichiometry at which the two substrates react with each other, or with the carrier, to form the penetrating complex. The ν values are negative for the species leaving, and positive for the species entering, a compartment.

Let us consider first a simple two-compartment system in which both coupled flows are osmotic. In accordance with the convention we shall denote the flows from compartment ' (left) to compartment " (right), as positive, and those in the opposite direction, as negative. The linked flows will now be treated like the "advancement of a chemical reaction". There is a certain arbitrariness as to which flow we should take as a reference flow, i.e., the flow representing the velocity of the overall reaction, J_r ($\nu = 1$). We could choose either J_a or J_b, but in accordance with the carrier concept, we shall take the flow of the complex between carrier and the two cosubstrates as the reference flow, i.e., as the rate of the quasi-chemical reaction, J_r. J_r has to be proportional to an affinity (A_r), L_r being the proportionality coefficient $J_r = L_r A_r$

The affinity of the overall reaction, A_r, is the sum of the negative electrochemical potential differences, each multiplied by the appropriate stoichiometric coefficient, as follows

$$A_r = -\Sigma \nu_i \, \mu_i$$

$$= -(\nu_a'' \, \mu_a'' + \nu_a' \, \mu_a' + \nu_b'' \, \mu_b'' + \nu_b' \, \mu_b') \qquad (7.10)$$

If both flows are osmotic, i.e., if neither species is chemically altered during the process, the ν referring to the same species are numerically equal on both sides:

$$|\nu_i'| = |\nu_i''| = \nu_i \qquad (7.11)$$

Hence we may replace ν_i' by $-\nu_i$ and ν_i'' by ν_i for the forward movement and vice versa for the backward movement. Accordingly we get

$$A_r = -(\nu_a \Delta \mu_a + \nu_b \Delta \mu_b) \qquad (7.12a)$$

or replacing

$$- \Delta \mu_i \text{ by } X_i$$

$$A_r = \nu_a X_a + \nu_b X_b \qquad (7.12b)$$

Obviously ν_b is positive in cotransport but negative in countertransport.

The rate of the overall reaction is

$$J_r = L_r A_r = L_r(\nu_a X_a + \nu_b X_b) \qquad (7.13)$$

and the rates of the indiviudal flows, to the extent that they are tightly coupled into the overall reaction, are

$$J_a = \nu_a J_r \text{ and } J_b = \nu_b J_r$$

or

$$J_a = \nu_a^2 L_r X_a + \nu_a \nu_b L_r X_b \qquad (7.14a)$$

$$J_b = \nu_a \nu_b L_r X_a + \nu_b^2 L_r X_b \qquad (7.14b)$$

7.1.4 Leakages. This notation, though superficially resembling the conventional notation of irreversible thermodynamics, differs from it in one important respect. It implies that the two species penetrate the membrane *only* jointly. This implication, as we shall see later, would mean that coupling is complete and the efficiency of energy transfer by this coupling is 100 %. Complete coupling, however, hardly ever occurs in reality and there are no membranes tight enough not to allow some leakage, for example, the free and uncoupled penetration of the transported species across pores or other leaky parts of the membrane.

This leakage we could call *outer leakage*. In addition, an *inner leakage* is also thinkable, which can be attributed to slipping of the carrier and to backcycling of the solute, as has been discussed before (p. 98 f).

To the extent that the leakage flows are proportional to the conjugate driving force of the leaking species, they should be

$$J_a^u = \Sigma L_{aa}^u \, X_a \tag{7.15a}$$

$$J_a^u = \Sigma L_{bb}^u \, X_b; \tag{7.15b}$$

the superscript u indicating that these flows are uncoupled. The summation sign is to indicate the possible existence of different parallel pathways of inner and outer leakage.

$$J_a = v_a \cdot J_r + \Sigma J_a^u \tag{7.16a}$$

$$J_b = v_b \cdot J_r + \Sigma J_b^u \tag{7.16b}$$

Except for special situations, which will be discussed later, only the straight terms, not the cross terms, need be corrected by leakage terms. Introducing the above leakage terms into Eqs. (7.14a and b) we obtain

$$J_a = (v_a^2 L_r + \Sigma L_{aa}^u) \, X_a + v_a v_b L_r X_b \tag{7.17a}$$

$$J_b = v_a v_b L_r X_a + (v_b^2 L_r + \Sigma L_{bb}^u) \, X_a \tag{7.17b}$$

The cross coefficients are still identical, as they should be according to Onsager's principles, but the two straight coefficients need no longer be equal for both species. These equations can be considered identical to the corresponding conventional equations derived before Eq. (7.6), but these equations are split into the coupled and the uncoupled or leakage components

$$L_{aa} = v_a^2 L_r + \Sigma L_{aa}^u \tag{7.18a}$$

$$L_{bb} = v_b^2 L_r + \Sigma L_{bb}^u \tag{7.18b}$$

$$L_{ab} = L_{ba} = v_a \cdot v_b L_r \tag{7.18c}$$

Owing to the leakage terms, as will be shown later, the coupling is no longer complete, and the efficiency of energy transfer is less than 100 %.

Since in negative coupling (e.g., in countertransport) v_b, is negative, also L_{ab}, the cross coefficient becomes negative. In a similar fashion we can handle chemi-osmotic coupling, e.g., between the flow of a species and the advancement of a chemical reaction in its simplest form:

$$\nu_s \ S \ \longleftrightarrow \ \nu_p \ P \qquad\qquad (7.19)$$

S and P stand for reactant and product, respectively.

In this case X_b is usually replaced by A_{ch}, the affinity of the chemical reaction involved. The index ch has been chosen to indicate a true chemical reaction, in contrast to the overall process, J_r, which is only operationally treated like a chemical reaction. As mentioned before, a true chemical reaction cannot simply be represented by the difference between the chemical potentials of the products and the reactants involved, because they may have different stoichiometric coefficients (ν). Accordingly $A_{ch} = -\Sigma \nu_i \mu_i$, or, applied to the simple reaction above,

$$A_{ch} = -(\nu_s \ \mu_s + \nu_p \ \mu_p). \qquad\qquad (7.20)$$

As before, the ν of the reactants are taken as negative and those of the products, as positive, according to general convention. In the quasi-chemical notation of this coupling we again have to consider leakage flows, here the uncoupled pathways of the reaction (S $\to$ P), the chemical equivalent of outer leakage.

The further procedure is analogous to that described above for osmo-osmotic coupling. The equations are:

$$J_a = (\nu_a^2 L_r + \Sigma L_{aa}^u) X_a + \nu_a L_r A_{ch} \qquad\qquad (7.21a)$$

$$J_{ch} = \nu_a L_r A_{ch} + (L_r + \Sigma L_{ch}^u) A_{ch} \qquad\qquad (7.21b)$$

We may again distinguish between outer and inner leakage. The outer leakage of the chemical part of the process has to account for the possibility that the chemical reaction may to some extent proceed without coupling with the flow of the above species, or may be coupled to another process. If, for instance, in a biological system A_{ch} stands for the affinity of the ATP hydrolysis, the outer leakage term for a given transport system directly coupled to ATP hydrolysis would include that part of ATP hydrolysis that is coupled to other processes that are also driven by the same ATP source. The equations look analogous to those derived for osmo-osmotic coupling.

Again, the cross coefficients are equal whereas the straight coefficients are expanded by leakage terms and hence unlikely to be the same for solute flow and chemical reaction rate. The reader may wonder why a stoichiometric coefficient is not being used for the chemical reaction, ν_{ch}. This coefficient will probably be unity in most cases and could be omitted. It might be used here to allow for the conceivable possibility that the hydrolysis of more than one ATP be required for every cycle.

7.1.5 Flux Ratio. It has been pointed out before that in free diffusion the flux ratio should be equal to the ratio of electrochemical activities of the solute species in question. This no longer holds in the presence of any kind of coupling, including tracer coupling. In the case of tracer coupling (isotope interaction), for instance, the flux ratio may also differ from the activity ratio, even in the absence of true coupling. Still, tracer coupling differs from true coupling, i.e., coupling to a nonconjugate flow, in that the flux ratio should become unity if the ratio of the electrochemical activities is also unity. Since this is not so with true coupling, a distinction between tracer and true coupling is in principle possible, to the extent that unity ratio of the activities can be verified experimentally and that the two fluxes can be measured under such conditions:

The relationship between flux ratio, driving forces, and tracer coupling can be briefly described as follows: Coupled transport between the transport of solute A and the flow of B, which could be the flux of another solute as in co- or countertransport or the advancement of a chemical reaction, can, in its simplest form, be described by the equation

$$J_a = L_{aa} X_a + L_{ab} X_b \tag{7.22}$$

or

$$X_a = R_{aa} J_a + R_{ab} J_b \tag{7.23}$$

According to Eq. (5.26) we have defined

$$RT \ln \boldsymbol{f}_a = J_a R_{aa}{}^* = \frac{J_a}{L_{aa}{}^*}$$

$R_{aa}{}^*$ and $L_{aa}{}^*$ referring to the "exchange resistance" or "exchange conductance", respectively. We get, by proper insertion:

$$RT \ln \boldsymbol{f}_a = \frac{R_{aa}{}^*}{R_{aa}} (X_a - R_{ab} J_b) \tag{7.25a}$$

or

$$RT \ln \boldsymbol{f}_a = \frac{L_{aa}}{L_{aa}{}^*} \left(X_a + \frac{L_{ab}}{L_{aa}} X_b \right) \tag{7.25b}$$

From previous discussions we know that

$$\frac{L_{ab}}{L_{aa}} X_b = \varepsilon_a X_b = RT \ln (HR) \tag{7.26}$$

ε_a being the "efficacy" of accumulation" of A, and HR, the Haldane ratio. We clearly see that at $X_a = 0$, i.e., if, the electrochemical activity ratio of A becomes unity, $RT \ln \boldsymbol{f}_a$ does not vanish here, in contrast to tracer coupling. At $X_a = 0$ we can say nothing about the true coupling unless we know the ratio $(R_{aa}{}^*/R_{aa})$ or $(L_{aa}/L_{aa}{}^*)$, for instance from the slope of $RT \ln \boldsymbol{f}_a$ if plotted as a function of X_a at constant X_b.

7.2 Mechanistic Correlates of the Phenomenological Parameters

7.2.1 General. As has been mentioned, the equations describing transport processes in terms of TIP can also be derived from the corresponding kinetic equations, previously developed in terms of the LMA. Since the latter ones are very much more involved, their "translation" into the simpler notation of TIP is very tedious unless some simplifying assumptions and approximations are introduced. Furthermore many kinetic paramters and their connection with the *concentrations* of the transported solute have to be lumped together to form the simple phenomenological coefficients of TIP. In this way much of the detailed information on the underlying mechanism is sacrificed. On the other hand this procedure allows us to visualize the energetic significance of detailed steps within the model and of their parameters. In particular we can in this way interpret inner leakage in terms of their mechanistic and kinetic correlates. Furthermore this procedure gives us the practically important possibility of dissecting the phenomenological coefficients into their kinetic constituents and thus to make reasonable estimates as to the range of applicability of the TIP notation.

An extensive study to relate the notation of transport processes in terms of TIP to that in terms of the LMA was carried out by KATCHALSKY and SPANGLER (1968) and by BLUMENTHAL and KEDEM (1969). The present treatment is based on a much more simplified model in order to make the fundamentals of the relationships better understandable.

The transformation of the involved equations based on the LMA into the simple ones of the TIP will be illustrated for some typical cases. To convert differences in concentrations or activities into the corresponding differences in electrochemical potentials we use the approximation of Eq. (5.17), assuming that the system operates within the range in which this approximation is valid. We shall see later, however, that similarly to the equations of facilitated diffusion, the range of applicability may be considerably extended for coupled systems by the choice of suitable conditions.

7.2.2 Secondary Active Transport. The conversion of an LMA equation into a TIP equation may first be illustrated for secondary active transport. For this purpose we start from Eq. (6.18) of the previous chapter: By introducing the above-mentioned approximation, we can replace all differences between the prime and the double prime values of each carrier species by the appropriate logarithmic expression, so that Eq. (6.18) becomes

$$J_a = \frac{x_T}{\Sigma} \left[rP_x P_{abx} \frac{\alpha'\beta' + \alpha''\beta''}{2} \ln\left(\frac{\alpha'}{\alpha''} \cdot \frac{\beta'}{\beta''}\right) \right.$$
$$\left. + (P_{ax}P_{bx} + rP_{bx}P_{abx}\beta'\beta'')\frac{\alpha' + \alpha''}{2}\ln\frac{\alpha'}{\alpha''} + P_{ax}P_{bx}\frac{\alpha'\beta'' + \alpha''\beta'}{2}\ln\frac{\alpha'}{\alpha''}\frac{\beta''}{\beta'} \right]$$

$$(7.27)$$

We now multiply numerator and denominator by *RT* separating the α-terms from the β-terms and introduce the *X* notation:

$$J_a = \frac{x_T}{RT\Sigma} \left\{ \left[r P_x P_{abx} \frac{\alpha'\beta'+\alpha''\beta''}{2} + (P_x P_{ax} + r P_{bx} P_{abx} \beta'\beta'') \frac{\alpha'+\alpha''}{2} \right. \right.$$
$$\left. + P_{ax} P_{bx} \frac{\alpha'\beta''+\alpha''\beta'}{2} \right] X_a$$
$$\left. + \left[r P_x P_{abx} \frac{\alpha'\beta'+\alpha''\beta''}{2} - P_{ax} P_{bx} \frac{\alpha'\beta''+\alpha''\beta'}{2} \right] X_b \right\} \quad (7.28)$$

The phenomenological coefficients are

$$L_{aa} = \frac{x_T}{RT\Sigma} \left[r P_x P_{abx} \frac{\alpha'\beta''+\alpha''\beta''}{2} + (P_x P_{ax} + r P_{bx} P_{abx} \beta'\beta'') \frac{\alpha'+\alpha''}{2} \right.$$
$$\left. + P_{ax} P_{bx} \frac{\alpha'\beta''+\alpha''\beta'}{2} \right] \quad (7.29a)$$

$$L_{ab} = \frac{x_T}{RT\Sigma} \left(r P_x P_{abx} \frac{\alpha'\beta'+\alpha''\beta''}{2} - P_{ax} P_{bx} \frac{\alpha'\beta''+\alpha''\beta'}{2} \right) \quad (7.29b)$$

Whether L_{ab} is positive, as in *cotransport*, or negative as in *countertransport*, depends obviously on the relative magnitudes of the two terms. In the quasi-chemical TIP treatment we have claimed that L_{ab} is equal to L_r, the overall reaction rate coefficient, and that L_{aa} is equal to L_r plus the leakage terms: To show this and also to find out the relationship between the inner leakage coefficients and the LMA parameters we simplify the above equation to make a pure cotransport model.

This can be accomplished by making either P_{ax} or P_{bx} equal to zero. We select all terms containing P_{ax} to be eliminated, which is tantamount to postulating either an "ordered reaction" in which the carrier combines with A only after binding of B (ax = o), or immobility of the carrier-A complex ($P_{ax} = o$). We obtain now for the phenomonelogical coefficients:

$$L_{aa} = \frac{r x_T}{RT\Sigma} \left(P_x P_{abx} \frac{\alpha'\beta'+\alpha''\beta''}{2} + P_{bx} P_{abx} \beta'\beta'' \frac{(\alpha'+\alpha'')}{2} \right) \quad (7.30a)$$

$$L_{ab} = \frac{r x_T}{RT\Sigma} \left(P_x P_{abx} \frac{\alpha'\beta'+\alpha''\beta''}{2} \right) \quad (7.30b)$$

If we set $L_{ab} = L_r$ we see that the inner leakage term is

$$L_{aa}^u = \frac{x_T}{RT} r P_{abx} P_{bx} \beta'\beta'' \frac{(\alpha'+\alpha')}{2} \text{ (slipping)} \quad (7.31)$$

L_{aa}^u refers to the inner leakage only since outer leakage is assumed to be absent from our model. The inner leakage is here related to the mobility of the binary complex XB and is therefore due to *slipping*.

Leakage by *backcycling* only is obtained analogously by the elimination of all terms with P_{bx}: The leakage coefficient is now

$$L_{aa}^u = \frac{x_T}{RT\Sigma} P_x P_{ax} \frac{\alpha'+\alpha''}{2} \text{ (backcycling)} \quad (7.32)$$

The analogous treatment can be applied to the pure countertransport model by setting either the terms with P_{abx} or those with P_x equal to zero. The terms of inner leakage will then be analogous to those of the cotransport model.

With the complete equation the quasi-chemical relationship is obscured, but still present: The overall flow can be interpreted to be composed of two flows, a cotransport-driven and a countertransport-driven one which have the leakage terms in common.

$$J_a = J_a^{\text{cotransport}} - J_a^{\text{countertransport}} \tag{7.33}$$

7.2.3 Primary Active Transport - Source-and-Sink Principle. The second example for the transformation of an LMA equation into a TIP equation will be a system of primary active transport of the source-and-sink type.

We have seen in the LMA treatment of such transport that the coupling to the driver reaction may be on the cis side $(Y \rightarrow X)$ or on the trans side $(X \rightarrow Y)$. With respect to transport in the cis-trans direction, in our notation from left to right, these couplings can be characterized as being of the *cis-coupled push* type and of the *trans-coupled pull* type, respectively. We have also demonstrated that by appropriately changing the affinities of the carrier species (X, Y) for the solute (A), and/or their mobilities, we can in either case invert the direction of the transport, i.e., from trans to cis, so that the *cis*-coupled push type becomes a *pull* type, and the *trans*- coupled pull type, a *push*- type coupling, with respect to the (inverted) transport direction. These relationships should also appear in the TIP treatment. To demonstrate this we shall now convert the LMA equations of primary transport into the corresponding TIP equations by a procedure analogous to that applied to secondary active transport.

We start from Eq. (6.42). By appropriately rearranging the numerator we obtain

$$J_a = \frac{x_T}{\Sigma} \left[rP_x P_{ax} \Gamma (\alpha' - \alpha'') + rP_y P_{ax} K_{xy} (\Gamma\alpha' - \alpha'') \right.$$
$$\left. + P_x P_{ay} K_{xy} (\alpha' - \Gamma\alpha'') + P_y P_{ay} K_{xy}^2 (\alpha' - \alpha'') \right] \tag{7.34}$$

Applying again the mentioned approximation to convert these differences into differences of chemical potentials we obtain

$$J_a = \left[\frac{x_T}{RT\Sigma} (rP_x P_{ax} \Gamma \frac{\alpha' + \alpha''}{2} + P_y P_{ay} K_{xy}^2 \frac{\alpha' + \alpha''}{2} + rP_y P_{ax} K_{xy} \frac{\Gamma\alpha' + \alpha''}{2} \right.$$
$$\left. + P_x P_{ay} K_{xy} \frac{\alpha' + \Gamma\alpha''}{2}) \right] X_a$$
$$+ \left[\frac{x_T}{RT\Sigma} (rP_y P_{ax} K_{xy} \frac{\Gamma\alpha' + \alpha''}{2} - P_x P_{ay} K_{xy} \frac{\alpha' + \Gamma\alpha''}{2}) \right] A_{\text{ch}} \tag{7.35}$$

setting $RT \ln \frac{a'}{a''} = X$ and $RT \ln \Gamma = A_{\text{ch}}$, the affinity of the (chemical) driver reaction.

As can easily be seen if this equation is written in the conventional phenomenological form

$$J_a = L_{aa} X_a + L_{ac} A_{ch},$$

the phenomenological coefficients being represented by the two expressions in brackets, the expression representing the cross coefficient, L_{ac}, consists of a positive and a negative term, which can be interpreted to represent a push term and a pull term, respectively. Hence the cross coefficient combines a positive and a negative coupling effect, as was also the case with the corresponding equations for secondary active transport (Eq. 7.29b). The positive term links the active transport in the forward direction, and the negative term the active transport in the reverse direction, to the nonconjugate driving force, A_{ch}. Since these two terms largely cancel each other, it can be presumed that in real systems, one of them will predominate, depending on the direction in which the active transport of the solute A is to take place.

For simplicity, we shall now reduce the equation to either a pure push-type pump or a pure pull-type pump. For this purpose we have to eliminate one of the opposing terms in the L_{ac}. It can be seen that the two terms, though both with positive signs, also appear on the expressions standing for L_{aa}, and whichever is to be eliminated in the cross coefficient, can also be eliminated in L_{aa}. We first start with the pure push-type pump, which transports in the forward direction only, so that we have to remove the pull-type term by making

$$r\, P_y P_{ax} K_{xy} \frac{\Gamma \alpha' + \alpha''}{2} \gg P_x P_{ay} K_{xy} \frac{\alpha' + \Gamma \alpha''}{2}$$

In other words, we assume that the complex AY is barely formed ($r \gg 1$) or is not translocated ($P_{ay} \to 0$). After eliminating this term in both L_{aa} and L_{ac}, we obtain the cross coefficient

$$L_{ac} = \frac{x_T}{RT\Sigma}\, r\, P_y P_{ax} K_{xy} \frac{\Gamma \alpha' + \alpha''}{2} \tag{7.36}$$

which in the quasi-chemical notation can be taken as being equal to L_r, i.e., the coefficient of the coupled overall process, and the straight coefficient

$$L_{aa} = \frac{x_T}{RT\Sigma}\, (r\, P_y P_{ax} K_{xy} \frac{\Gamma \alpha' + \alpha''}{2} + r\, P_x P_{ax} K_{xy} \Gamma \frac{\alpha' + \alpha''}{2} + P_y P_{ay} K_{xy}^2 \frac{\alpha' + \alpha''}{2})$$

$$\tag{7.37}$$

Writing this term in the quasi-chemical notation, we see that it contains L_r and two additional terms that can be identified as leakage terms. Hence $L_{aa} = L_r + \Sigma\, L_{aa}^u$. The first leakage term

$$\frac{x_T}{RT\Sigma}\, r\, P_x P_{ax} \Gamma \frac{\alpha' + \alpha''}{2}$$

represents *slipping*, whose significance is largely determined by P_x, the translocation probability of the empty carrier. The other

$$\frac{x_T}{RT\Sigma} \, P_y \, P_{ay} \, K_{xy}^2 \, \frac{\alpha'+\alpha''}{2}$$

represents *backcycling*, which is largely determined by P_{ay}, the translocation probability of the loaded inactive carrier.

Complete inversion of the transport direction with the same coupling (cis side) is obtained by removing the *push* terms. For this purpose we assume that

$$r \, P_y \, P_{ax} \frac{\Gamma\alpha'+\alpha''}{2} \ll P_x \, P_{ay} \frac{\alpha'+\alpha''\Gamma}{2} \tag{7.38}$$

as might be achieved by preventing the formation of complex AX ($r \ll 1$) or by immobilizing it ($P_a \to 0$). The cross coefficient now reduces to the "pull" term and thus becomes negative. In the quasi-chemical notation it is numerically equal to the overall rate coefficient.

$$L_r = - \, L_{ac} = \frac{x_T}{RT\Sigma} \, P_x \, P_{ay} \, K_{xy} \frac{\alpha'+\Gamma\alpha''}{2} \tag{7.39}$$

The leakage terms are the same as in Eq. (7.37), but with an inverted meaning: The first one, with P_{ax}, now represents backcycling, and the second one, with P_y, slipping.

We see again that with the same model the direction of active transport can be inverted by appropriate changes of affinities of the carrier species for the solute and of the velocities of the various carrier species'.

The analogous treatment can be applied also to the model with the trans side coupling, but will be omitted here, since it does not supply new information.

It is possible to develop TIP equations for the transport of solute A by using instead of X_b or A_{ch}, respectively:

$RT \ln (HR)$

as the nonconjugate driving effort. The resulting rate equation would have the same rate coefficient (L_{aa}) for both conjugate and nonconjugate driving efforts, since the leakages, which are responsible for the differences between the phenomenological coefficients, are already implicit in HR, the Haldane Ratio (p. 100)

$$J_a = L_{aa} \, (X_a + RT \ln (HR)) \tag{7.40}$$

At $J_a = 0$ one would obtain directly the maximum static head. Otherwise the equation is different from the phenomenological TIP equation and at present does not appear to have special advantages. It is seen that

$$RT \ln (HR) = - \, \varepsilon_a \, (X_b) \tag{7.41}$$

ε_a being the efficacy of accumulation, (L_{ab}/L_{aa}).

7.2.4 Comparison Between LMA Treatment and TIP Treatment. The equations
of secondary and primary active transport derived on the basis
of the TIP obviously are much simpler and shorter than the corresponding equations and terms of the LMA, but they say almost
nothing about the underlying mechanism. Accordingly all models
of primary active transport can be expressed by identical equations formally, except that the phenomenological coefficients
will of course have different numerical values for the different
models. The same is true for secondary active transport: Also
here the equations look formally the same, regardless of whether
they are of the affinity type, or mixed type model. Moreover,
they do not allow a distinction with respect to the driving gradient, i.e., whether it is more chemical or more electrical in
nature, since the driving force X_i combines both driving forces
in a single value, even though each of these driving force components may affect the overall kinetics in a different way.

If we want to decide which of two alternative mechanistic models
would best fit a given process, the equations of irreversible
thermodynamics are of no immediate help. On the other hand, they
may be very suitable for providing us with other important information, especially concerning the energetics of the mechanism.
On the basis of suitable experiments, they may, for instance, tel
us which energy sources are involved in driving a certain process
At least they may show us whether the energy sources anticipated
are adequate. Of particular interest in many cases is the degree
of coupling (q). It not only tells us how tight the coupling between two linked processes is, it may also help us to distinguish
between two possible sources of energy, i.e., to determine which
one is coupled directly to the transport process under investigation, and which one is not coupled at all or at least very indirectly. This kind of analysis has been applied successfully,
e.g., to show that the active transport of certain amino acids
in certain cells is driven immediately by an electrochemical potential gradient of Na ions rather than directly by a chemical
metabolic reaction (HEINZ and GECK, 1974). The same principle
would also be helpful to decide on analogous alternatives in othe
transport systems, for instance in the transport of organic solutes in microorganisms. Finally the equation of TIP, if properly
derived, may tell us something about the efficacy by which the
energy supplied by a certain sources is utilized for a certain
work. The solution of these and other energetic questions may
be promoted by suitable application of the principles of TIP.

It appears to be very instructive to relate the phenomenological
TIP equations to the corresponding kinetic (LMA) equations. Where
as the equations in terms of the TIP do not tell us how the corresponding equations in terms of the LMA would look, the reverse
is in principle possible. Accordingly, a given equation of a
coupled transport process derived on the basis of the LMA can
be approximately transformed into the corresponding equation in
terms of the TIP, as we have done with a few examples. Such a
transformation gives us some idea how and where the various parameters of the kinetic equation are incorporated into the phenomenological TIP coefficients. Obviously, each phenomenological
coefficient is a very involved function, not only of a great number of different parameters but also of the activities of the

solutes involved. Hence they are not true constants but can still
be treated as such with sufficient approximation under suitable
conditions. The analysis of each single phenomenological coefficient as to the constituting parameters and activities can tell
us also something about these conditions and thus about the presumable limits set to the application of TIP equations to real
transport systems.

It is interesting to compare the treatment of the various kinds
of leakages in the quasi-chemical equations of TIP with that of
the same leakages in the LMA equations. In our quasi-chemical
treatment of coupling we have derived the straight coefficient
simply by adding the leakage coefficients, both inner and outer
ones, to the overall rate coefficient (L_r), which was assumed to
be identical with the cross coefficients. By converting LMA equations into TIP equations we have learned that this simple procedure of incorporating inner leakages is under certain conditions
justified. We could verify the expected relationship between the
leakage terms and those LMA parameters that are responsible for
the carrier-mediated dissipation of energy. We have also seen,
however, that there are limitations to the simple procedure,
namely, that if both slipping and recycling are appreciable at
the same time, part of the transport is inverted, which upsets
the quasi-chemical relationship. Hence the simple handling of
internal leakage in the quasi-chemical procedure requires that
the product of the coefficients of the slipping and of the recycling, respectively, is negligible. A priori, we are inclined
to presume that in real transport systems this requirement is
fulfilled, since a system actively transporting in two opposite
directions at the same time would be almost prohibitively inef-
ficient and wasteful. Plausible ways to shun inner leakage pathways would be, for primary active transport, to reduce the affinity of the "inactive" form of the carrier for the transportable solute, and/or to immobilize the complex between the inactive carrier and the transportable solute.

In cotransport an analogous affinity effect may come about by
an "ordered" reaction sequence, in that, for instance, the cosolute has to be bound by the carrier prior to the driven solute.
In countertransport, the opposite must be true, i.e., the driver
solute must first be removed before the driven solute is bound,
and vice versa. Hence it appears plausible to use a single term
for inner leakage and to disregard transport inversion effects
in biological transport systems so that the simple quasi-chemical
procedure suffices to derive the phenomenological coefficients
from the overall reaction rate coefficient. However, the experimental proof has still to be provided.

On the other hand, the close interrelationship between forward
and backward transport within the same model provokes some interesting speculations: Theoretically it should be possible to shift
a given active transport system into the reverse direction, simply by modulating the relative magnitudes of affinity and/or velocity parameters of the model. The question arises whether there
are transport systems in which nature makes use of such a possibility.

7.3 Quantitative Evaluation of Coupling

7.3.1 Degree of Coupling. The coupling between flows may be more or less "tight". As a quantitative measure of this tightness the degree of coupling (q) has been introduced. q can be positive or negative, the sign indicating whether the coupling is positive, as in cotransport, or negative, as in countertransport. The numerical value of q varies between 0 and 1, the former indicating absence, and the latter, completeness of coupling. q is a function defined in terms of parameters of irreversible thermodynamics:

$$q = \frac{L_{ij}}{\sqrt{L_{ii} \cdot L_{jj}}} = -\frac{R_{ij}}{\sqrt{R_{ii} \cdot R_{jj}}} \qquad (7.42)$$

In positive coupling L_{ij} is positive, in negative coupling, negative. The converse is true for R_{ij}.

In the quasi-chemical notation, which explicitly distinguishes between coupled and uncoupled (leakage) flows

$$q = \frac{\nu_i \nu_j L_r}{\sqrt{(\nu_i^2 L_r + \Sigma L_{ii}^u)(\nu_j^2 L_r + \Sigma L_{jj}^u)}} \qquad (7.43)$$

q becomes negative if either one of the stoichiometric coefficients (ν_i or ν_j) is negative, as in negative coupling (e.g., in countertransport).

One sees that q depends heavily on the magnitude of the leakage coefficients relative to that of the overall reaction coefficient (L_r). In the absence of inner and outer leakge q would be 1 (or -1 in negative coupling). With increasing leakage q becomes numerically smaller and approaches zero if the sum of leakage coefficients is so great as to make L_r negligible.

Any driven process may be coupled directly to one, and indirectly to another driving process. For instance, the flow of an actively transported species may be directly coupled to the splitting of an energy-rich phosphate, and indirectly, to O_2 consumption. Since continuous transport depends on the resynthesis of ATP, which in turns is coupled to respiration, the transport process will ultimately also be coupled to the O_2 consumption of the system. The coupling between transport and O_2 consumption will have a smaller q than that between the transport and the ATP hydrolysis. The overall q of indirect coupling should equal the product of the q values of all single steps linked in series:

$$q_{overall} = q_1 \cdot q_2 \cdots \cdot q_n \qquad (7.44)$$

If several processes are energetically coupled in series their order might be obtained from a comparison of the q values.

q will be the larger the closer the coupled reactions are to each other in the sequence.

The practical significance of the degree of coupling will now be illustrated for the simple model described before (Fig. 25), in which the flow of solute A is coupled to the downhill flow of solute B.

q is experimentally accessible through the determination of the mutual dependence between the coupled flows (stoichiometric numbers). So far the coupling between the flow of species A and that of the species B,

$$q^2 = \left(\frac{\partial J_a}{\partial J_b}\right)_{X_a} \cdot \left(\frac{\partial J_b}{\partial J_a}\right)_{X_b} , \text{ or } = \frac{(X_a)_{J_a=0}}{(X_a)_{J_b=0}} \qquad (7.45)$$

A practical procedure would be to determine first the stoichiometric number of A versus B, i.e., the increment of the flow of A, J_A, As a function of an increase in the flow of B, J_B, at constant X_a, then the analogous stoichiometric number of B versus A, and to multiply the obtained partial differentials by one another, according to the first of the above equations.

7.3.2 Efficiency of Coupling — Efficacy of Accumulation. The q value may be used to determine the efficiency (η) of a transport system, i.e., the negative ratio of power output to power input (KEDEM and CAPLAN, 1965).

$$\eta = - \frac{J_a X_a}{J_b X_b} \qquad (7.46)$$

Whereas q may be constant for a transport system with constant parameters under all conditions, η varies with the functional state of the system as with the transport rate. The maximum efficiency (η_{max}), however, is a function of q only and hence constant as long as q is constant (KEDEM and CAPLAN, 1965):

$$\eta_{max} = \frac{q^2}{(1 + \sqrt{1 - q^2})^2} \qquad (7.47)$$

We see from the equation that η_{max} approaches unity in complete coupling, but falls more rapidly than does q.

Since η_{max} is a function of q only, it does not give any additional information about the system. It does not allow prediction, for instance, of the highest static head accumulation ratio or of the highest level flow transport of a given system under defined conditions. A more useful parameter for this purpose might be the *efficacy of accumulation* (ε_a) defined as the maximum ratio of accumulation potential to the driving potential at static head

$$\varepsilon_a = \left(\frac{X_a}{X_b}\right)_{J_a=0} \qquad \text{for secondary active transport} \qquad (7.48a)$$

$$\varepsilon_a = \left(\frac{X_a}{A_{ch}}\right)_{J_a=0} \qquad \text{for primary active transport} \qquad (7.48b)$$

For a two-flow system

$$\varepsilon_a = \frac{\nu_a \, \nu_b \, L_r}{\nu_a \, \nu_b \, L_r + \Sigma L_{aa}^u} = \frac{L_{ab}}{L_{aa}} = \left(\frac{\partial J_b}{\partial J_a}\right)_{X_b} \tag{7.49}$$

We see that the stoichiometry of the coupling need not be known to determine ε_a.

ε_a is in some ways analogous to the "reflection coefficient" (σ) used in the quantitative treatment of the coupling between solvent flow and solute flow through aqueous pores, and to the "heat of transfer" (Q^*) used in the treatment of coupling between water and heat flow. ε_a is also related to the Haldane ratio by the expression

$$\varepsilon_a = \frac{HR}{X_b} \tag{7.50}$$

7.3.3 Power of Transport Systems. Transport systems are often characterized as more or less "powerful". The question arises, by which objective and measurable standards should this "powerfulness" be assessed? One could, in analogy to technical mashines, give for this purpose the maximum power that a given transport system, perhaps per unity dry weight or other standard of reference, is able to put out. In other words, one could determine the greatest possible product of net transport rate of a given solute and its conjugate electrochemical potential gradient. However, even the optimal value that one may measure will certainly be smaller than the intrinsic, or ideal, power of the system, i.e., the power that could be put out in the complete absence of inner and outer leakages. As we have discussed before, part of the leakage occurs via parallel pathways by which the accumulated solute may escape irreversibly down its electrochemical gradient (outer leakage), and the other part concerns the transport system itself (inner leakage). These leakages would have to be known if the intrinsic power of the system were to be assessed. In most cases it is very difficult, or even impossible, to appreciate the extent of these leakages and hence to get an idea of the intrinsic power of the transport system under consideration.

The power concept fails with systems that are in a state which comes close to either *static head*, or *level flow*, respectively, since in either case the overall power is zero.

In *static head* a considerable concentration or electrochemical potential gradient may be maintained, but the measurable net movement is zero, such as is the case in *homeocellular* transport. The power of such a system would seemingly be zero though it is certain that considerable osmotic work is expended. To assess the true energy output of such a system, one would have to separate the active and passive flows in *parallel*, and to determine their rates, which at static head are opposite and equal. Either of these flows, multiplied by the static head electrical potential difference, would give the numerical minimum power output of the system (HEINZ and MARIANI, 1957).

In *level flow* the electrochemical potential gradient is virtually,
or close to, zero, so that again, despite extensive transport
with the expenditure of energy, the effective power output is
zero. Such a situation may occur in many cases of *transcellular*
transport, e.g., in the intestine or in the kidney tubules. To
assess the true power output, one would have to separate the
passive and active steps *in series*, and to determine their electrochemical potential differences, which should be equal, but in
opposite directions. Either of these electrochemical potential
differences multiplied by the level flow rate would give the minimum numerical power output of the system. These dissections of
transport systems into parallel pathways or into pathways in
series are often difficult or even impossible.

A useful parameter for characterizing the effectiveness of an
active transport system in either static head or level flow is

RT ln (HR),

HR being the Haldane ratio, which, as shown before, should be
the same if determined in either state. As we pointed out previously, this value represents the maximum available nonconjugate
driving force and may in some cases be verified experimentally
in the steady state. It is equal to εX_b, i.e., the idel nonconjugate driving force (X_b) times the efficacy of accumulation ε,
as defined before. Hence RT ln (HR) does not give the *intrinsic*
power, but only the available power of the system, i.e., the
maximum power of the transport in the absence of outer leakage.

Apart from these uncertainties the power of the transport system,
even if it can be measured experimentally, may not always be satisfactory as an energetic characterization of a transport system. It is, for instance, not very discriminatory between two
kinds of systems, which apart from their power, are distinctly
different and appear to serve different functions: On the one
hand, there are systems that may develop and maintain high electrochemical potential differences but have only a small transport
capacity, i.e., the maximum transport rate is small under all
circumstances. Such systems are, for instance, involved in the
maintenance of unequal distributions of solutes between different
compartments, e.g., between cells and medium. On the other hand,
there are systems that are unable to produce an appreciable electrochemical potential gradient, but may transport large amounts
of solutes across biological membranes. Such systems are at work
in the absorption of solutes by intestine and kidney tubules. It
is clear that the characteristic difference between these transport systems does not appear in their power output: For this
purpose some surrogate standars rather than the true power output
have been suggested, such as the "efficacy of force" for static
head or the "efficacy of flow", for level flow, dividing simply
the maximum static head accumulation ratio for the maximal rate
of level flow, respectively, by the appropriate energy input
(ESSIG and CAPLAN, 1968). Clearly the units of these parameters,
time per mole for static head, and moles per energy for level
flow, are very unusual and may raise objections against the use
of these parameters. Still, they may be helpful if a distinction
between the two kinds of transport systems is desired.

7.4 Limitations of Thermodynamics of Irreversible Processes (TIP) Treatment

7.4.1 Proximity to Equilibrium. The notation of irreversible thermodynamics implies linearity between flows and driving forces. The restricted validity of this assumption severely limits the applicability of the phenomenological equations. As has been pointe out before, the implied linearity is an approximation that is useful only to the extent that the deviation from reality is small enough not to exceed experimental error or, at least, not to upset the conclusions drawn from the treatment of processes in terms of irreversible thermodynamics. To the extent that the rates of processes obey the law of mass action, i.e., as long as these rates are linear functions of concentrations or activities, the same rates cannot be linear functions of thermodynamic potential differences as well. The latter may be treated as such only within a narrow region near equilibrium limited by the condition that the logarithm of the concentration ratio, for instance, must not appreciably exceed the value of the first term of the appropriate series, as given in Eq. (5.17)

$$\ln \frac{c_i'}{c_i''} \approx 2 \, \frac{c_i' - c_i''}{c_i' + c_i''} + \ldots$$

This is true between $c_i' = 3 \, c_i''$ and $c_i' = \dfrac{c_i''}{3}$.

This range corresponds to about 1 RT, i.e., about 0.6 kcal/mol, and within this range the deviation from linearity does not exceed 10 %, provided, however, that the arithmetic means of the two activities remain constant throughout.

Experimentally, this latter condition is difficult to meet, if only in one of the two compartments the activity of the transported solute can be varied at will, for instance, in studies with cellular or subcellular systems. In such cases one usually tries to keep activity of the transported solute constant in one compartment and to vary it in the other. As a consequence the deviation from linearity of the curves obtained is increased and the 10 % limit of the applicability of the TIP treatment is at about 0.2 RT (Fig. 31), i.e., the total driving force, A_r, must be within ± 0.2 RT. For a single process, this is indeed a very narrow range, a range almost so small as to prohibit a good deal of practical biological application. Still, the situation may not be quite as hopeless for many coupled systems in the steady state under two conditions. Firstly, in coupled systems the above limitation applies to the "global" driving force, i.e., to the affinity of the overall reaction rather than to the partial driving forces of the individual substances involved. In such systems the various partial driving forces that make up the global affinity have in the steady state opposite signs, so that the global affinity may be considerably smaller than most of the partial forces per se. Let us take for example the Na/K pump in the cellular membrane. In the steady state each K and Na is maintained at considerable potential differences, each of which would by far exceed the applicability of irreversible ther-

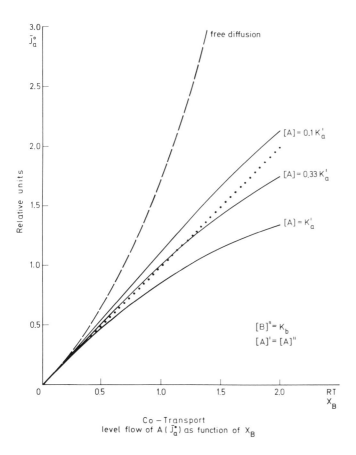

Fig. 31. Linearity of TIP equations in cotransport. The net transport of solute A is plotted against the chemical potential difference of cosolute B, X_b. The concentration of B is kept constant on both sides of the membrane at the value indicated. The concentration of the cosolute B is kept constant on the trans side at the value of K_b; whereas the concentration of B on the cis side is varied in order to vary X_B. The *broken line* would obtain if A passes the membrane exclusively by free diffusion, without cotransport. The *solid lines* would obtain if the transport of A were achieved exclusively by cotransport with B according to the equations of the mass action law in the text. The units of X_A are in RT, the product of gas constant times absolute temperature

modynamics. Since they have, however, opposite signs as compared to the immediate driving force, let us say, ATP hydrolysis, the overall difference in thermodynamic potential is likely to be much smaller, and hence more favorable to the treatment in terms of irreversible thermodynamics. Since, however, this overall A_r may still be greater than 0.2 RT, it would be advisable to test first for efficient linearity by possibly tedious procedures.

7.4.2 Other Quasi-Linear Ranges. Secondly, as we have seen with the
TIP treatment of facilitated diffusion, saturable systems offer
the possibility to expand considerably the range of quasi-linearity of the flow-driving force relation, if the right concentration range is selected. We have shown before (see Fig. 10)
that with facilitated diffusion linearity may extend beyond ± 1
kcal, if one of the solute concentrations is kept at about K_m
and the other varied around this value. For coupled transport
such possibilities have not been investigated experimentally,
but for cotransport the equations derived on the basis of the
LMA predict a likewise extended range of linearity that is best
if each of the concentrations A and B have values in a given
relation to the standard parameters. Domains of linearity are
not only restricted to the range near equilibrium with saturable
systems. Even far away from equilibrium ($4 - 5$ RT) linear regions
may exist. ROTTENBERG (1973) developed some ingenious mathematically simple operations to detect such domains experimentally
and to use them to study the process in terms of TIP. These operations are exemplified at an enzymatic reaction in which a substrate (S) is converted to its product (P): One possibility would
be to keep P fixed at a given value that can be obtained from the
two Haldane relationships and vary S around K_m, the Michaelis
constant of the forward reaction. The reaction rate, plotted
against the logarithm of the relative concentration of S, i.e.,
against the logarithm of S divided by its Michaelis constant

$$\log \frac{S}{K_S}$$

follows over a considerable range a straight line passing through
the origin. This linearity is almost perfect between the concentrations $S = \frac{K_S}{2}$ and $S = 2\,K_S$, which corresponds to a range of
about 1.4 RT or of about 1 kcal/mol. If one is satisfied by a
deviation of less than 10 %, the region of quasi-linearity may
cover almost 2 kilocalories. Another possibility is to correct
both the flow of the reaction and the driving force by half the
maximum rate. Again, one obtains a true linear range of a similar
order of magnitude as above. The final possibility is to keep S
constant at a value that can also be derived from the Haldane
relationship and to vary the concentration of P between 2S and
1/2S. Again, linearity is satisfactory over a range similar to
that given above. All these possibilities require that the reaction so treated is far from equilibrium, in particular, that the
maximum reaction rates for the forward and backward reaction,
respectively, are very different from each other.

Even though these derivations are carried out for an enzymatic
reaction, they clearly should also hold for active transport
processes so that also these could be treated in terms of irreversible thermodynamics with sufficient accuracy under conditions
far remote from equilibrium that should easily be within the
range of experimental verification.

Outside the ranges of quasi-linearity, the application of irreversible thermodynamics may become very complicated and thereby
lose its conveniences. It can be anticipated that the single

driving forces, X_i and A_r, no longer appears solely in their first power. Terms may appear which contain, for instance, X_i^2, A_r^2 or XA_r etc. For practical purposes, it may therefore be preferable to stay within quasi-linear regions, i.e., either close to equilibrium or, in special cases, within other ranges of quasi-linearity, such as are likely to exist and to be verified in particular with saturable reactions in the range of half saturation, such as have been described above and which are numerous among biological processes.

8 Phase-Specific Forces

In the previously treated systems the driving forces were solute-specific, i.e., they were arranged so as to affect more or less specifically the transport of a single solute species. These driving forces originated from electrochemical potential gradients or from affinities of reactions that were stoichiometrically coupled to the transport system, usually via a solute-specific carrier mechanism.

There are also driving forces that do not act on a single solute species but on a compartment or phase as a whole and could therefore be called "phase-specific". They may affect the flow of many if not all permeant components of the system. In other words, there may be coupling between the fluxes of some solute species on the one hand, and the flow of solvent, in most cases water, or the flow of electricity, or the flow of heat, on the other hand. Hydrostatic and/or osmotic pressure gradients, electrical potential gradients, and temperature gradients belong in the group of phase-specific forces. An electric potential gradient, however, can be considered phase-specific only to a limited extent since its direct effect is on permeant ions only.

To the extent that phase-specific forces can discharge only via solute flow, they may be treated as components of the conjugate forces for any such solute involved. This is true primarily for the electrical PD, which is therefore treated as part of the electrochemical potential difference of all ions present. Accordingly it is not customary to speak of coupling between the flow of matter and the flow of electricity in the case of ionic migration. The electrical PD will therefore not be treated as a phase-specific force in the present context.

Typical phase-specific forces are differences in hydrostatic (or osmotic) pressure and in temperature. As indicated in the Introduction, both forces are of minor significance as conjugate forces for solute flows because they are too small. They may, however, become significant with respect to solvent (water) flow and may thus have an indirect effect on solute flows. Their effect on water flow will be treated in the next section.

8.1 Drag Forces Between Solute and Solvent Flows

8.1.1 Mechanism. In the foregoing, any movement of solvent (H_2O) and its possible effect on the solute flows have been disregarded, because our models, chosen for illustration, have been devised so as to keep the flow of solvent negligible. In reality, however, the flow of solvent particles can often not be avoided. The flow of solvent is likely to interact with the various solute flows through the membrane, especially if they use the same pathways. By this interaction the flows of both solvent and solute will be modulated and some phenomena of coupling will be observed. The extent of these interactions depends on the nature of the membrane.

A membrane that transmits only water, without any solute movement accompanying it, is called "semipermeable". On the other hand, a membrane that is equally permeable for the solvent and all solutes present alike is called "nonselective". In neither kind of membrane should interaction between solutes and solvent be significant. This is different for "permselective" membranes, which transmit water and solutes to a different extent.

Most biological membranes, such as the cellular membranes, are neither semipermeable nor completely nonselective, but are more or less permselective, i.e., they transmit only certain solutes and may further discriminate between these according to their size, shape, electric charge or chemical properties, and they usually transmit water more easily than most solutes.

Before dealing with the solute-solvent interactions we shall first have a look at the flow of solvents in general, in particular at that of water (KOEFOED-JOHNSON and USSING, 1953).

The movement of water through membranes is to some extent analogous to that of solutes; however, it shows pecularities that follow from the continuous and "quasi-crystalline" arrangement of solvent molecules in fluids. Accordingly, one distinguishes between "diffusional flow" and "bulk flow" of solvent through membranes. In *diffusional* flow each solvent molecule has to be detached from the bulk in order to penetrate the membrane phase by "free diffusion", like a solute particle. Diffusional water flow is best exemplified by the diffusion of evaporated water molecules through a gas phase. The process may be similar if water molecules pass through a nonporous lipid layer, but it is doubtful whether diffusional water flow across lipid areas of biological membrane, if it occurs at all, is quantitatively relevant except, as we shall see later, in the presence of temperature gradient.

In *bulk flow*, on the other hand, the solvent molecules are assumed to retain their continuity owing to strong cohesive forces. The so-called quasi-crystalline structure of water in the bulk phase may be greatly distorted during the process of penetrating a pore, but each water molecule remains in immediate contact with other water molecules of its immediate neighborhood. Bulk flow predominates if water moves through pores or through the mesh-work of nonlipid constituents of a barrier, as possibly in cell membranes. Unlike diffusional flow, which follows Fick's diffusional law, bulk flow supposedly follows Poiseuille's law of fluid streaming across narrow channels. Through artificial nonlipid porous membranes, such as collodium or cellophane membranes, 98 - 99 % of the movement of water is by bulk flow, regardless of whether a hydrostatic or osmotic pressure difference is the driving force. In pure lipid membranes, which do not contain pores, bulk flow should not occur at all, but is supposedly difficult to exclude

in the presence of so-called unstirred layers maintained close
to the interfaces of the membrane. In biological membranes, which
are assumed to consist of both lipid and possibly hydrophilic
porous nonlipid components, such as proteins, etc., buth bulk
and diffusional flow are believed to occur. According to previous
calculations, bulk flow appears to predominate, but owing to unstirred layers near the membrane interfaces, may be overestimated
Hence, at present, it does not appear possible to assess quantita
tively the contribution of each mode of the overall movement of
H_2O through biological membranes. The great ease, however, at
which H_2O penetrates cell membranes, as compared to the rather
slow penetration of most solutes, suggests that bulk flow of water accounts for an appreciable part of water permeation.

During bulk flow through pore-like structures, frictional interaction between solute and solvent is likely to be considerable
and may have important consequences for the behavior of both: It
appears that solute and solvent "drag" each other along across
the membrane. For example, the flow of water through pores or
through some water-filled meshwork, induced by a difference in
hydrostatic or osmotic pressure, may cause solutes to flow agains
their electrochemical potential gradient; vice versa, the movemen
of solutes down their concentration gradient across a membrane
may drag water against the osmotic pressure gradient, i.e., from
higher to lower osmolarity. The latter effect appears to have
special biological significance whenever much water is being
transported, for instance, in intestinal or renal secretion and
reabsorption. It is, however, not customary to call these drag
effects active transport, even though molecules of solute or solvent, respectively, may be moved against their electrochemical
potential gradient, probably because the interaction involved is
not stoichiometric, i.e., the number of solvent molecules moving
per solute particle is not fixed but varies with the circumstances, depending on concentration, membrane structure, etc.
Through the coupling between the flow of a solute and that of
the solvent, hydrostatic pressure differences between biological
compartments, too small to affect the activities of solutes directly, may have significant effects on solute translocation.

8.1.2 A Model. The following model was the first suggested to explain this water movement and its linkage to active ion transport,
without the assumption of a special molecular transport system
for water molecules (HEINZ, 1960). We need a compartment, B,
which is separated from another compartment, A, by a semipermeable membrane. A solute is actively pumped into B from A by a
mechanism located possibly in only a limited area of the separating membrane, which has otherwise rather rigid walls. Compartment B is connected through one or more channels or holes with
a third compartment, C, into which the "transport" of water is
to take place. Obviously, the active transport of a solute from
A into B will raise the osmotic presssure in B so that water will
flow from A to B through the membrane, which for this purpose
has to be "semipermeable", i.e., easily permeable to water but
poorly permeable, or even impermeable, to the passively moving
solutes. One may assume that this behavior is due to the fact
that the membrane has pores that are wide enough to let water
pass but too narrow for the diffusion of the solute concerned.

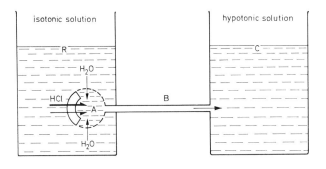

isotonic solution

hypotonic solution

Fig. 32. Model of water movement against osmotic gradient (description in text)

This flow of water will cause an increase in the hydrostatic pressure in B, especially if the membrane is fairly rigid, and as a consequence bulk flow of solution will take place through the channels or holes into the final compartment, C (Fig. 32). For the proper function of this model the dimensions of the channels or holes between B and C have to meet two requirements: Firstly, they must be wide enough to prevent osmotic water flow from C to B. Secondly, they must be narrow and long enough to permit the maintenance of a hydrostatic pressure gradient between B and C to continuously drive the solvent across. Under suitable conditions, a steady state is conceivable in which continuous net transport of solute into B will cause a continuous outflow of solution, ultimately from B into C, even if C ocntains distilled water. Obviously, it is not pure water that flows from B into C but rather a solution whose osmolarity is even somewhat higher than that in B and C. C, however, may be so large that its osmolarity is little affected for a long time, during which time water is moved, in essence, in the direction from higher (A) to lower osmolarity (C) in other words, against its thermodynamic driving force. This model has been modified by CURRAN (1965), who replaced the long channel (S) by a membrane with pores wide enough not to discriminate between solvent and solute, but, on the other hand, tight enough to maintain the required hydrostatic pressure difference.

It has been shown experimentally that a device manufactured accordingly and fitted by suitable membranes of differential permeability (reflection coefficient) indeed displayed seemingly active movement of water if solute was added to the middle compartment to mimic active solute transport. The histology of water-pumping tissues appears to be such as to have a biological equivalent to the above model structure. The compartment in question would be based between two cells (DIAMOND and TORMEY, 1966). It is known that this compartment is fairly wide but narrows down considerably at the basal end of the cellular layer so that the channel is being formed that corresponds to the channel in our model.

The quantitative treatment of drag forces between the flows of a solute and the solvent cannot simply be carried out formally in terms of the LMA, such as we deal with solute-specific coupling in secondary active transport. In the latter case the coupling comes about by the formation of specific and stoichiometric

complexes between the species concerned, directly or mediated
by a carrier. Clearly solute-solvent interactions is by friction
and hence neither stoichiometric nor specific. Furthermore, the
treatment of solute-solute interaction in secondary active transport has been simplified by the reasonable assumption that the
penetration occurs by jumps so that the formation and dissociation of transportable complexes could be restricted to the interfaces. Solute-solvent interaction, however, is probably continuous along aqueous pores or the water-filled meshes or filamentous
regions, so that, for instance, the ratio between the penetrating
solute particles and the co-migrating solvent molecules is likely
to vary during the passage.

Attempts have been made to incorporate the drag forces into the
conjugate force of the solute under consideration, similar to
the incorporation of electromotive forces into the conjugate
driving force of ions. A hydrostatic or osmotic pressure difference would then correspond to the electric PD. This procedure,
however, though it is appropriate for handling the electromotive
force, appears to be less suitable to handle drag forces. It
should be applied only if the force in question discharges only
via the flow of the solute species into whose conjugate driving
force it is to be incorporated, so that the same phenomenological
coefficient holds for both the chemical and the electric PD. This
is certainly different with drag forces. A hydrostatic pressure
difference can discharge via volume flow but its effect on solute
flow depends on various circumstances, such as friction coefficients, etc.

8.1.3 Treatment of Terms of TIP. Nowadays the more elegant treatment
in terms of irreversible thermodynamics, going back to KEDEM and
KATCHALSKY (1958), is usually preferred. For illustration we
choose again a model consisting of two compartments, ' and ",
and a rigid permselective membrane that is differentially permeable to solute and solvent, but, in contrast to previous models,
offers no restraint to water movement. To restrict ourselves to
two components we assume that only one solute species (S) is present, water (W) being the second component. We assume that solute
S is initially present in both compartments at the concentrations
c_S' and c_S'', respectively. The flow of water and that of the solute,
each multiplied by the corresponding conjugate force, will add
to the entropy production ($\dot{S}$)

$$T\dot{S} = -(J_S \Delta \mu_S + J_W \Delta \mu_W) \tag{8.1}$$

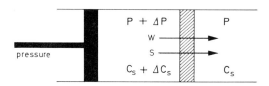

Fig. 33. Solute-solvent drag, simplified model. Coupling between the movement
of solute (*S*) and water (*W*) across a membrane under the influence of differences in hydrostatic pressure (Δ*P*) and concentration differences of solute
(Δ*C*$_S$) between the two compartments

J_S and J_W are the flows of solute and solvent, respectively, per unit membrane area (Fig. 33). In the presence of a pressure difference (ΔP) the two potential differences can be split into two terms based on partial differentials, one at constant pressure and the other at constant concentration:

$$\Delta \mu_S = \left(\frac{\partial \mu_S}{\partial c_S}\right)_P \Delta c_S + \left(\frac{\partial \mu_S}{\partial P}\right)_{c_S} \Delta P = \Delta \mu_S^C + \overline{V}_S \Delta P \qquad (8.2a)$$

$$\Delta \mu_W = \left(\frac{\partial \mu_W}{\partial c_W}\right)_P \Delta c_W + \left(\frac{\partial \mu_W}{\partial P}\right)_{c_W} \Delta P = \Delta \mu_W^C + \overline{V}_W \Delta P \qquad (8.2b)$$

Assuming that the chemical potentials of S and W do not directly affect each other,

$$\left(\frac{\partial \mu_S}{\partial c_W}\right)_{c_S} = \left(\frac{\partial \mu_W}{\partial c_S}\right)_{c_W} = 0,$$

hence

$$T\dot{S} = - J_S (\Delta \mu_S^C + \overline{V}_S \Delta P) - J_W (\Delta \mu_W^C + \overline{V}_W \Delta P) \qquad (8.3)$$

$\overline{V}_S$ and $\overline{V}_W$ are the partial molar volumes of solute and water, respectively. For the terms at constant pressure the Gibbs-Duhem equation can be applied

$$n_S d\mu_S^C + n_W d\mu_W^C = 0 \qquad (8.4)$$

or, dividing by V', the volume of compartment '

$$c_S d\mu_S^C + c_W d\mu_W^C = 0$$

or $dc_S + dc_W = 0$

as $d\mu_S^C = RT \frac{dc_S}{c_S}$ and $d\mu_W^C = RT \frac{dc_W}{c_W}$.

Under the condition that the concentration difference between the compartments is small enough, we may with sufficient approximation replace the differentials by the concentration differences, Δc_S and Δc_W, between the compartments, so that

$$\Delta c_S = - \Delta c_W .$$

Since $\qquad\qquad\qquad\qquad\qquad\qquad\qquad\qquad\qquad\qquad (8.5)$

$$\Delta \mu_S^C = \frac{RT}{c_S} \Delta c_S \quad \text{and} \quad \Delta \mu_W^C = \frac{RT}{c_W} \Delta c_W = - \frac{RT}{c_W} \Delta c_S$$

Eq. (8.3) finally becomes

$$T\dot{S} = - \left[J_S RT \frac{\Delta c_S}{c_S} + J_S \overline{V}_S \Delta P - J_W RT \frac{\Delta c_S}{c_W} + J_W \overline{V}_S \Delta P \right]$$

or, after rearrangement

$$T\dot{S} = - (J_s \bar{V}_s + J_w \bar{V}_w) \Delta P - \left(\frac{J_s}{c_s} - \frac{J_w}{c_w}\right) RT\Delta c_s \tag{8.6}$$

$(J_s \bar{V}_s + J_w \bar{V}_w)$ obviously is the volume flow, J_v; $\left(\frac{J_s}{c_s} - \frac{J_w}{c_w}\right)$ is the exchange flow, J_D, i.e., the solute flow relative to the water flow.

J_D is a peculiar flow that calls for some interpretation. It represents the excess solute flow, i.e., the flow in excess of that associated with J_v, if the flowing solution had the solute concentration of the cis compartment (c_s'). The extra solute flow is expressed in volume units, i.e., as the volume required to make the concentration of the extra solute transported equal to c_s'. If, for instance, any flow of S is accompanied by the appropriate flow of water, $J_w = \frac{c_w}{c_s} J_s$

$$\frac{J_s}{c_s} = \frac{J_w}{c_w} \tag{8.7}$$

i.e., J_D is zero; as would be the case with a completely non-selective membrane.

Hence each term represents a flow of energy, but in contrast to previous equations of this kind, each energy flow is expressed here as the product of volume flow and pressure difference rather than of molar flow and electrochemical potential difference.

From the entropy equation we can now derive the phenomenological equations:

$$J_v = L_p \Delta P + L_{pD} RT\Delta c_s \tag{8.8}$$

$$J_D = L_{Dp} \Delta P + L_D RT\Delta c_s \tag{8.9}$$

L_p, L_{pD}, L_{Dp}, and L_D are the phenomenological coefficients. According to Onsager's principle we can assume that

$$L_{pD} = L_{Dp} \tag{8.10}$$

If no other solute is present, $RT\Delta c_s$ is equal to the difference in osmotic pressure

$$RT\Delta c_s = \Delta \pi_s \tag{8.11}$$

and at static head ($J_v = 0$)

$$\Delta P = - \frac{L_{pD}}{L_p} \Delta \pi_s \tag{8.12a}$$

$$\Delta P = - \sigma_s \Delta \pi_s; \tag{8.12b}$$

σ is called the reflection coefficient. It can be used as an index of the permselectivity of the membrane.

If σ in the same system were unity, no solute could pass and

$$J_D = \frac{J_w}{c_w} = - L_p \Delta P$$

i.e., the system behaves as if with each iso-osmotic volume transported, the amount of S in this volume were pushed back into the
original compartment.

σ_s is unity if the membrane is ideally semipermeable — being permeable to water only, not to the solute. σ_s is zero if the membrane does not discriminate between water and solute at all. Intermediate values of σ_s ($0 < \sigma_s < 1$) give the degree of leakiness of
the membrane to the solute. σ may be negative if the membrane is
more permeable to the solute than to the solvent.

As has been mentioned before, σ_s has a meaning analogous to ε_a,
the efficacy of accumulation of a transport mechanism, since $\sigma_s =$
(L_{pD}/L_p) and $\varepsilon_a = \cdot$ (L_{ab}/L_a).

8.2   Thermo-Osmotic Coupling

8.2.1 Mechanism. The coupling between the flow of matter and that
of heat (entropy) across a membrane is a well-established physical fact. It is doubtful, however, or even unlikely that this
kind of coupling is significant in biological transport. Most
biological systems are considered isothermal, and there is good
reason to believe that biological membranes, such as the cellular
membranes, are too thin to maintain sizable temperature differences for any appreciable length of time. Still, it cannot be
completely excluded that exothermic metabolic processes within
the cells may locally give rise to small transient temperature
gradients. Owing to thermo-osmotic coupling, a temperature gradient may cause a flow of matter, e.g., solutes or solvent,
through the membrane, or conversely an electrochemical potential
difference of solutes, or the solvent may cause a flow of heat
through the membrane. In either case a transduction of energy
occurs, of heat energy into osmotic energy and vice versa.

As we recall from previous discussions, coupling requires that
the flows concerned use a joint pathway. The flow of matter and
that of heat may, for instance, do so through a membrane that
transmits only activated particles, i.e., particles furnished
with the required "energy of activation". Such a membrane functions as a special "energy barrier", which is overcome only by
those particles that are "hotter" than the rest. Each of these
"activated" particles passing the membrane carries an extra amount
of heat from one compartment to the other. Consequently the flow
of matter, for instance of a gas down an imposed pressure gradient, may eventually build up to a temperature difference between the two compartments, provided the membrane is sufficiently
insulated against nonparticulate heat conductance. Conversely,
the flow of heat down an imposed temperature gradient through the
same membrane may eventually build up a hydrostatic pressure difference between the compartments. To apply such "thermo-osmotic"
coupling to a biological membrane, we choose as an example the

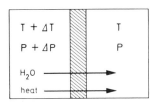

Fig. 34. Thermo-osmosis. Coupling between the flow of H_2O and heat. ΔT the difference in temperature and ΔP the difference in hydrostatic pressure between the two adjacent solutions. For details see text

permeation of water between two compartments through a membrane that transmits water mainly by diffusional flow. In other words, each water molecule, in order to pass the membrane via this pathway, has to be evaporated from the bulk phase, as if the membrane were a gas space. For simplicity, we assume that solutes, whenever present in the bulk phases, cannot pass the membrane. If we raise the temperature and/or the hydrostatic pressure in the left compartment relative to the right one, as illustrated in Figure 34, flows of heat and matter (water) will ensue from left to right through the membrane, which are to some extent thermo-osmotically coupled to each other. The mutual interaction between the flow of matter and that of heat can be advantageously treated in terms of irreversible thermodynamics. The underlying principle of this treatment will be given here, but in view of the uncertainty as to its biological significance, only briefly.

8.2.2 Treatment in Terms of TIP. According to Onsager's phenomenological approach we state first the energy dissipation by the two coupled flows in this system, i.e., the rate of entropy production times the absolute temperature $(T\dot{S})$. The entropy production in the present system is composed of the entropy production by heat flow down a temperature gradient $(\dot{S}_Q)$ and of that by the flow of water down its electrochemical potential gradient $(\dot{S}_w)$. The former is equal to the heat flow times the difference in inverse temperature between the two compartments

$$\dot{S}_Q = J_Q \Delta \frac{1}{T} = - J_Q \frac{\Delta T}{T^2} \tag{8.13}$$

and the latter is equal to the molecular flow of water times the difference in the partial molar entropy of water between the two compartments

$$\dot{S}_w = J_w \Delta \bar{S}_w \tag{8.14}$$

Since according to the second law of thermodynamics $T\bar{S}_w = \bar{H}_w - \mu_w$, $\bar{H}_w$ being the partial molar enhalpy, and μ_w the chemical potential of water, we get

$$\dot{S}_w = J_w \left[\Delta \left(\frac{\bar{H}_w}{T} \right) - \Delta \left(\frac{\mu_w}{T} \right) \right] \tag{8.15}$$

from which we get by differentiation, provided that the differences are small enough and that $\Delta \bar{H}_w = 0$:

$$\dot{S}_w = - J_w \frac{\bar{H}_w \Delta T + T \Delta \mu_w - \mu_w \Delta T}{T^2} \tag{8.16}$$

According to the second law of thermodynamics

$$\mu_w = \bar{H}_w - T\bar{S}_w$$

and

$$n_w \Delta \mu_w = V \Delta P - S \Delta T$$

or

$$\Delta \mu_w = \bar{V}_w \Delta P - \bar{S}_w \Delta T$$

or in the presence of an osmotic pressure difference $(\Delta \pi_s)$

$$\Delta \mu_w = \bar{V}_w (\Delta P - \Delta \pi_s) - \bar{S}_w \Delta T$$

inserting this into Eq. (8.16), we get

$$\dot{S}_w = - J_w \frac{\bar{V}_w (\Delta P - \Delta \pi_s)}{T} \tag{8.17}$$

and for the total dissipation of the system:

$$T\dot{S} = - \left[J_Q \frac{\Delta T}{T} + J_w \bar{V}_w (\Delta P - \Delta \pi_s) \right] \tag{8.18}$$

from which we derive the phenomenological equations

$$J_Q = - \left[L_{QQ} \frac{\Delta T}{T} + L_{Qw} \bar{V}_w (\Delta P - \Delta \pi_s) \right] \tag{8.19}$$

$$J_w = - \left[L_{wQ} \frac{\Delta T}{T} + L_{ww} \bar{V}_w (\Delta P - \Delta \pi_s) \right] \tag{8.20}$$

$$L_{Qw} = L_{wQ}$$

At $\Delta T = 0$

$$\frac{J_Q}{J_w} = \frac{L_{Qw}}{L_{ww}} = Q^* \tag{8.21}$$

Q^* is the heat of transfer, i.e., the heat that accompanies the flow of water under isothermal conditions. In other words, in order to prevent temperature differences during the flow of water, we would have to add the amount Q^* of heat to the left compartment and to withdraw the same amount of heat from the right compartment for each mol of water moving through the membrane. If we replace L_{Qw} by $L_{ww} Q^*$ in the equation of the water flow we get

$$J_w = - \left[Q^* L_{ww} \frac{\Delta T}{T} + L_{ww} \bar{V}_w (\Delta P - \Delta \pi_s) \right] \tag{8.22}$$

It is seen that the water flow stops ($J_w = 0$) if

$$\frac{\Delta P - \Delta \pi}{\Delta T} = - \frac{Q^*}{V_w T} \tag{8.23}$$

In the absence of a hydrostatic pressure an osmotic pressure dif-

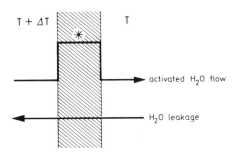

$T + \Delta T$ T

activated H$_2$O flow

H$_2$O leakage

Fig. 35. Thermo-osmosis, interpreted as activated H$_2$O flow across
the barrier. For details see text

ference equivalent to $\frac{Q^*}{\bar{V}_w T}$ (thermo-osmotic pressure) can be maintained per degree of temperature difference. Q^* depends on the
nature of the membrane; it approximately equals the energy required to remove one mol of water from the bulk solution into
the membrane phase. To get an idea of the magnitude we follow a
rough calculation of Spanner's based on the assumption that Q^*
refers to an "aereal" membrane, i.e., about equivalent to the hea
of evaporation, i.e., about 10,476 cal/mol = 432.90 liter-atmos-
pheres. With pure lipid membranes, Q^* is likely to be even highe
than that (SPANNER, 1954).

Since $\bar{V}_w$ is about 0.018 liter we get

$$\frac{\Delta \pi_s}{\Delta T} = \frac{Q^*}{\bar{V}_w T} = \frac{432.90}{0.018 \cdot 298} = 80.7 \text{ atm/grad}$$

This means that the osmotic pressure must be 80 atm higher on
the warmer compartment in order to balance a flow of water at a
temperature difference of one degree. We see that even a very
small difference in temperature may cause water to move against
a considerable osmotic gradient.

It may be that the temperature differences between animal cells
and their natural environment are too small to give rise to significant thermo-osmotic effects, but this may not be so with
other systems, for instance plant cells. It is noteworthy that
while the flow of water is stopped if the thermo-osmotic pressure
has reached its critical value, the flow of heat may continue.
This is because zero flow of water refers to the total movement
of water, i.e., if the thermo-osmotic water flow is balanced by
uncoupled (leakage) water flow through other channels (Fig. 35).
Hence, to maintain a "static head" in osmotic or hydrostatic
pressure, the heat-coupled water flow must continue and so must
the flow of that associated with it. The situation is similar to
the accumulation of solute by a pump-and-leak system.

References

BLUMENTHAL, R., KATCHALSKY, A.: The effect of the carrier association-disso-
ciation rate on membrane permeation. Biochim. Biophys. Acta 173, 357 (1969)

BLUMENTHAL, R., KEDEM, O.: Flux ratio and driving forces in a model of active
transport. Biophys. J. 9, 432 (1969)

van BRUGGEN, J.T., BOYETT, J.D., van BUEREN, A.L., GALEY, W.R.: Solute flux
coupling in a homopore membrane. J. Gen. Physiol. 63, 639 (1974)

CHIPPERFIELD, A.-R., WHITTAM, R.: Evidence that ATP is hydrolysed in a one-
step reaction of the sodium pump. Proc. R. Soc. London A 187, 269 (1974)

CURRAN, P.F.: Ion transport in intestine and its coupling to other transport
processes. Fed. Proc. 24, 993 (1965)

DAINTY, J., HOUSE, C.R.: An examination of the evidence for membrane pores
in frog skin. J. Physiol. 185, 172 (1966)

DIAMOND, J.M., TORMEY, J.McD.: Role of long extracellular channels in fluid
transport across epithelia. Nature (London) 210, 817 (1966)

DIAMOND, J.M., WRIGHT, E.M.: Biological Membranes: The physical basis of ion
and nonelectrolyte selectivity. Ann. Rev. Physiol. 31, 581 (1969)

EDWARDS, P.A.W.: Evidence for the carrier model of transport from the inhibition by NEM of cholin transport across the human red cell membrane. Biochim.
Biophys. Acta 307, 415 (1973)

EDWARDS, P.A.W.: A test for non-specific diffusion steps in transport across
cell membranes, and its application to red cell glucose transport. Biochim.
Biophys. Acta 345, 373 (1974)

EHRENSTEIN, G., LECAR, H.: The mechanism of signal transmission on nerve
axons. Ann. Rev. Biophys. Bioeng. 1, 347 (1972)

EISENMANN, G.: On the elementary atomic origin of equilibrium ionic specificity. In: Membrane Transport and Metabolism (eds. A. KLEINZELLER, A. KOTYK),
p. 608. New York: Academic Press, 1961

EISENMANN, G.: Cation selective glass electrodes and their mode of operation.
Biophys. J. 2, 259 (1975)

ESSIG, A., CAPLAN, S.R.: Energetics of active transport processes. Biophys.
J. 8, 1434 (1968)

FOSTER, D.M., JACQUEZ, J.A.: An analysis of the adequacy of the asymmetric
carrier model for sugar transport. Biochim. Biophys. Acta 436, 210 (1976)

GECK, P.: Eigenschaften eines asymmetrischen Carrier Modells für den Zucker-
transport am menschlichen Erythrozyten. Biochim. Biophys. Acta 241, 462
(1971)

GECK, P., HEINZ, E.: Coupling in secondary transport - Effect of electrical
potentials on the kinetics of ion-linked co-transport. Biochim. Biophys.
Acta 443, 49 (1976).

HALDANE, J.B.C.: Enzymes. London: Longmans, Green, 1930

HALL, J.E., MEAD, C.A., SZABO, G.: A barrier model for current flow in lipid
bilayer membranes. J. Membr. Biol. 11, 75 (1973)

HASSELBACH, W.: Sarcoplasmic membrane ATPase. In: The Enzymes, Vol. X, p. 431.
New York: Academic Press, 1974

HECKMANN, K.: Zur Theorie der "Singele File"-Diffusion I. Z. Physik. Chem.
44, 184 (1964)

HEINZ, E.: Kinetic studies on the influx of glycine-1-C^{14} into the Ehrlich
mouse ascites carcinoma cell. J. Biol. Chem. 211, 781 (1954)

HEINZ, E.: Grundmechanismus der Magensäureproduktion und deren Regulation. Klin. Physiol. 1, 184 (1960)

HEINZ, E.: Transport of amino acids by animal cells. In: Metabolic Pathways (ed. L. HOKIN), Vol. VI, p. 455. New York: Academic Press, 1972a

HEINZ, E.: Na-Linked Transport of Organic Solutes (ed. E. HEINZ). Berlin-Heidelberg-New York: Springer, 1972b

HEINZ, E.: Coupling and energy transfer in active amino acid transport. In: Current Topics in Membranes and Transport (eds. F. BRONNER, A. KLEINZELLER), Vol. IV, p. 137. New York: Academic Press, 1974

HEINZ, E., DURBIN, R.P.: Studies of the chloride transport in the gastric mucosa of the frog. J. Gen. Physiol. 41, 101 (1957)

HEINZ, E., GECK, P.: The efficiency of energetic coupling between Ng^+ flow and amino acid transport in Ehrlic cells - a revised assessment. Biochim. Biophys. Acta 339, 426 (1974)

HEINZ, E., GECK, P.: The electrical potential difference as a driving force in Na-linked cotransport of organic solutes. In: Coupled Transport Phenomena in Cells and Tissues. The Peter F. Curran Memorial Symposium (ed. J.F. HOFFMAN), Vol. I, p. 13. New York: Raven Press, 1977

HEINZ, E., GECK, P., WILBRANDT, W.: Coupling in secondary active transport. Activation of transport by co-transport and/or counter-transport with the fluxes of other solutes. Biochim. Biophys. Acta 255, 442 (1972)

HEINZ, E., MARIANI, H.A.: Concentration work and energy dissipation in active transport of glycine into carcinoma cells. J. Biol. Chem. 228, 97 (1957)

HEINZ, E., WALSH, P.M.: Exchange diffusion, transport and intracellular level of glycine and related compounds. J. Biol. Chem. 233, 1488 (1958)

JACQUEZ, J.A.: The kinetics of carrier-mediated active transport of amino acids. Proc. Nat. Acad. Sci. U.S.A. 47, 153 (1961)

JARDETZKY, O., SNELL, F.M.: Theoretical analysis of transport processes in living systems. Proc. Nat. Acad. Sci. U.S.A. 46, 616 (1960)

KABACK, H.R.: The role of the phosphoenolpyruvate-phosphotransferase system in the transport of sugars by isolated membrane preparations of escherichia coli. J. Biol. Chem. 248, 3711 (1968)

KATCHALSKY, A., CURRAN, P.F.: Nonequilibrium Thermodynamics in Biophysics. Cambridge, Mass.: Harvard University Press, 1965

KATCHALSKY, A., SPANGLER, R.: Circulation due to concentration gradients: facilitated transport. In: Dynamics of Membrane Processes, Quarterly Review of Biophysics I. p. 127, 1968

KEDEM, O.: Criteria of active transport. In: Membrane Transport and Metabolism (eds. A. KLEINZELLER, A. KOTYK), p. 87. London-New York: Academic Press, 1960

KEDEM, O., CAPLAN, S.R.: Degree of coupling and its relation to efficiency of energy conversion. Trans. Faraday Soc. 61, 1897 (1965)

KEDEM, O., ESSIG, A.: Isotope flows and flux ratios in biological membranes. J. Gen. Physiol. 48, 1047 (1965)

KEDEM, O., KATCHALSKY, A.: Thermodynamic analysis of the permeability of biological membranes non-electrolytes. Biochim. Biophys. Acta 27, 229 (1958)

KLINGENBERG, M., RICCIO, P., AQUILA, H., BUCHANAN, B.B., GREBE, K.: Mechanism of carrier transport and the ADP, ATP carrier. In: The Structural Basis of Membrane Function (eds. Y. HATEFI, L. DJAVADI-OHANIANC), p. 293. New York-London: Academic Press, 1976

KOEFOED-JOHNSEN, V., USSING, H.H.: The contribution of diffusion and flow to the passage of D_2O through living membranes. Acta Physiol. Scand. 28, 60 (1953)

LeFEVRE, P.G.: A model for erythrocyte sugar transport based on substrate-conditioned "Introversion" of binding sites. J. Membr. Biol. 11, 1 (1973)

LIEB, W.R., STEIN, W.D.: The molecular basis of simple diffusion within biological membranes. In: Current Topics in Membranes and Transport (eds. F. BRONNER, A.KLEINZELLER), Vol. II, p. 1. New York: Academic Press, 1971

LIEB, W.R., STEIN, W.D.: Carrier and non-carrier models for sugar transport in the human red blood cell. Biochim. Biophys. Acta 265, 187 (1972)

MAHLER, H.R., CORDES, E.H.: Biological Chemistry. New York: Harper and Row, 1971

MARTONOSI, A.: Transport of calcium by the sarcoplasmic reticulum. In: Metabolic Pathways, Vol. VI, p. 317. New York: Academic Press, 1972

MEISTER, A.: On the enzymology of amino acid transport. Science 180, 33 (1973)

MILLER, D.M.: The kinetics of selective biological transport. III. Erythrocyte-monosaccharide transport data. Biophys. J. 8, 1329 (1968a)

MILLER, D.M.: The kinetics of selective biological transport. IV. Assessment of three carrier systems using the erythrocyte-monosaccharide transport data. Biophys. J. 8, 1339 (1968b)

MILLER, D.M.: The kinetics of selective biological transport. V. Further data on the erythrocyte-monosaccharide transport system. Biophys. J. 11, 915 (1971)

MILLER, D.M.: The effect of unstirred layers on the measurement of transport rates in individual cells. Biochim. Biophys. Acta 266, 85 (1972)

MITCHELL, P.: Biological transport phenomena and the spatially anisotropic characteristics of enzyme systems causing a vector component of metabolism. In: Membrane Transport and Metabolism (eds. A. KLEINZELLER, A. KOTYK), p. 22. London-New York: Academic Press, 1960

MITCHELL, P.: Coupling of phosphorylation to electron and hydrogen transfer by a chemiosmotic type of mechanism. Nature (London) 191, 144 (1961)

MITCHELL, P., MOYLE, J.: Group-translocation: A consequence of enzymecatalysed group-transfer. Nature (London) 182, 372 (1958)

MULLINS, L.J.: The penetration of some cations into muscle. J. Gen. Physiol. 42, 817 (1959)

MULLINS, L.J.: Ion selectivity of carriers and channels. Biophys. J. 15, 921 (1975)

PAGANELLI, C.V., SOLOMON, A.K.: The rate of exchange of tritiated water across the human red cell membrane. J. Gen. Physiol. 41, 259 (1957)

PATLAK, C.S.: Contributions to the theory of active transport. II. The gate type non-carrier mechanism and generalizations concerning tracer flow, efficiency, and measurement of energy expenditure. Bull. Math. Biophys. 19, 209 (1957)

PELLEFIGUE, F., DeBROHUN BUTLER, J., SPIELBERG, St.P., HOLLENBERG, M.D., GOODMAN, St.I., SCHULMAN, J.D.: Normal amino acid uptake by cultured human fobrablasts does not require gamma-glutamyl transpeptidase. Biochem. Biophys. Res. Com. 73, 997 (1976)

POST, R.L., KUME, S., ROGERS, F.N.: Alternating paths of phosphorylation of the sodium and potassium ion pump of plasma membranes. In: Mechanisms in Bioenergetics (eds. G.F. AZZONE, L. ERNSTER7),. p. 203. New York-London: Academic Press, 1973

POST, R.L., MERRIT, C.R., KINSOLVING, C.R., ALBRIGHT, C.D.: Membrane adenosine triphosphatase as a participant in the active transport of sodium and potassium in the human erythrocyte. J. Biol. Chem. 235, 1796 (1960)

POSTMA, P.W., ROSEMAN, S.: The bacterial phosphoenol pyruvate: sugar phosphotransferase system. Biochim. Biophys. Acta 457, 213 (1976)

RING, K., HEINZ, E.: Active amino acid transport in streptomyces hydrogenans. I. Kinetics of uptake of α-aminoisobutyric acid. Biochem. Z. 344, 446 (1966)

ROSENBERG, T., WILBRANDT, W.: Uphill transport induced by counter flow. J. Gen. Physiol. 41, 289 (1957)

ROTTENBERG, H.: The thermodynamic description of enzyme-catalyzed reactions. Biophys. J. <u>13</u>, 503 (1973)

SCHATZMANN, H.: ATP-dependent Ca-extrusion in the human red cell. Experientia <u>22</u>, 364 (1966)

SCHULMAN, J.D., GOODMAN, S.I., MACE, J.W., PATRICK, A.D., TIETZE, F., BUTLER, E.J.: Gluthathionuria: inborn error of metabolism due to tissue deficiency of gamma glutamyl transpeptidase. Biochem. Biophys. Res. Com. <u>65</u>, 68 (1975)

SEN, A.K., WIDDAS, W.F.: Determination of the temperature and pH dependence. J. Physiol. <u>160</u>, 392 (1962)

SINGER, S.J.: Membrane Structure and Transport. In: Abstracts (6. Ann. ICN-UCLA Symposia, Keystone, Colorado) <u>1</u>, 120 (1977)

SINGER, S.J., NICOLSON, G.L.: The fluid mosaic model of the structire of cell membranes are viewed as two-dimensional solutions of oriented globular proteins and lipids. Science <u>175</u>, 720 (1972)

SOLOMON, A.K.: Measurement of the equivalent pore radius in cell membranes. In: Membrane Transport and Metabolism (eds. A. KLEINZELLER, A. KOTYK), p. 94. New York-London: Academic Press, 1960

SPANNER, D.C.: The active transport of water under temperature gradients. Symposia Soc. Exp. Biol., Cambridge, Univ. Press <u>8</u>, 76 (1954)

SPANNER, D.C.: Introduction to Thermodynamics. London-New York: Academic Press, 1964

STEIN, W.D.: The Movement of Molecules across Cell Membranes. New York: Academic Press, 1967

STEIN, W.D., LIEB, W.R., KARLISH, S.J.D., EILAM, Y.: A model for active transport of sodium and potassium ions as mediated by a tetrameric enzyme. Proc. Nat. Acad. Sci. U.S.A. <u>70</u>, 275 (1973)

USSING, H.H.: The distinction by means of tracers between active transport and diffusion. Acta Physiol. Scand. <u>19</u>, 43 (1949)

USSING, H.H.: Some aspects of the application of tracers in permeability studies. In: Advances in Enzymology and Related Subjects of Biochemistry (ed. F.F. NORD), Vol. VIII, p. 21. New York: Interscience, 1952

VIDAVER, G.A.: Inhibition of parallel flux and argumentation of counter flux shown by transport models not involving a mobile carrier. J. Theor. Biol. <u>10</u>, 301 (1966)

WHITTAM, R., CHIPPERFIELD, A.R.: The reaction mechanism of the sodium pump. Biochim. Biophys. Acta <u>415</u>, 149 (1975)

WILBRANDT, W.: Zuckertransporte. In: Biochemie des aktiven Transports (12. Coll. d. Ges. f. Biol. Chemie), p. 112. Berlin-Göttingen-Heidelberg: Springer, 1961

WILBRANDT, W., ROSENBERG, T.: The concept of carrier transport and its corollaries in pharmacology. Pharmacol. Rev. <u>13</u>, 109 (1961)

WILSON, T.H., KUSCH, M., KASHKET, E.R.: A mutant in *escherichia coli* energy-uncoupled for lactose transport: A defect in the lactose-operon. Biochem. Biophys. Res. Com. <u>40</u>, 1409 (1970)

Subject Index

Molecular Biology, Biochemistry and Biophysics

Editors: A. Kleinzeller, G.F. Springer,
H.G. Wittmann

Springer-Verlag
Berlin
Heidelberg
New York